Lecture Notes in Electrical Engineering

Volume 350

About this Series

"Lecture Notes in Electrical Engineering (LNEE)" is a book series which reports the latest research and developments in Electrical Engineering, namely:

- Communication, Networks, and Information Theory
- Computer Engineering
- Signal, Image, Speech and Information Processing
- Circuits and Systems
- Bioengineering

LNEE publishes authored monographs and contributed volumes which present cutting edge research information as well as new perspectives on classical fields, while maintaining Springer's high standards of academic excellence. Also considered for publication are lecture materials, proceedings, and other related materials of exceptionally high quality and interest. The subject matter should be original and timely, reporting the latest research and developments in all areas of electrical engineering.

The audience for the books in LNEE consists of advanced level students, researchers, and industry professionals working at the forefront of their fields. Much like Springer's other Lecture Notes series, LNEE will be distributed through Springer's print and electronic publishing channels.

More information about this series at http://www.springer.com/series/7818

Vinod Pangracious · Zied Marrakchi
Habib Mehrez

Three-Dimensional Design Methodologies for Tree-based FPGA Architecture

Springer

Vinod Pangracious
Electrical and Computer Engineering Department, School of Engineering
American University in Dubai
Dubai
United Arab Emirates

Habib Mehrez
University of Pierre and Marie Curie, Paris VI
Paris
France

Zied Marrakchi
Flexras Technologies
Biocitech
Romainville
France

ISSN 1876-1100 ISSN 1876-1119 (electronic)
Lecture Notes in Electrical Engineering
ISBN 978-3-319-19173-7 ISBN 978-3-319-19174-4 (eBook)
DOI 10.1007/978-3-319-19174-4

Library of Congress Control Number: 2015941877

Springer Cham Heidelberg New York Dordrecht London

Printed on acid-free paper

Springer International Publishing AG Switzerland is part of Springer Science+Business Media (www.springer.com)

To my wife Juliet Vinod, and my kids Liz, Lia and Yehoshua

Preface

Three-dimensional integrated circuits (3D-ICs) design has become a major driving force in the modern VLSI design and manufacturing technology. It provides the best platform for design and manufacturing of high density and high performance chips and the field is continuing to grow at an amazing pace. The idea to write this book on 3D FPGAs using 3D technology originated from the research and experimental work done during my doctoral studies at University of Pierre and Marie Curie under the guidance of Professor Habib Mehrez. This book was written as a text that covers the foundations of 3D integrated circuits and high performance 3D reconfigurable FPGA architecture design. It was written for use in a core and elective course at the graduate level in field of Electrical Engineering, Computer Engineering, and doctoral research programs. Today, many universities upgrade their curriculum to include modern VLSI design methodologies and re-configurable system design. No previous background on 3D integration is required, nevertheless, fundamental understanding of 2D CMOS VLSI design is required. It is assumed that the reader has taken the core curriculum in Electrical Engineering or Computer Engineering, with courses like CMOS VLSI design, Digital System Design and Microelectronics Circuits being the most important. It is accessible for self-study by both senior students and professionals alike.

Scope and Coverage

A brief introduction and the concept of 3D integration is presented in Chap. 1. It begins with brief review of advanced VLSI design and technology scaling; Chap. 1 continues to establish the basic and fundamental needs of introducing three-dimensional integrated circuits design into the modern VLSI technology. It also stress the needs for new and augmented 3D CAD tools to support designs such as, the *design for 3D*, to manufacture high performance 3D integrated systems and reconfigurable architecture. Three-dimensional (3D) integration is an emerging technology that is expected to lead to an industry paradigm shift due to its tremendous advantages over 2D integration in terms of density and performance. Academic and industrial research institutes around the world currently focus on technology innovations, simulation and design and product prototypes. Anticipated applications start with memory, portable device and high-performance computers, reconfigurable system design and extend to high-density multifunctional heterogeneous integration of infotech-nanotech-biotech systems. Chapter 2 focuss on the fundamentals and in-depth analysis of different 3D integration methodologies and design. This chapter also talks about potential benefits of 3D integration that can vary depending on approach; they include multi-functionality, increased performance, reduced power, small form factor, reduced packaging, increased yield and reliability, flexible heterogeneous integration and reduced overall costs.

Today, FPGAs (Field Programmable Gate Arrays) have become important actors in the computational devices domain that was originally dominated by microprocessors and ASICs. The main challenge in 2D FPGA design is to find a good trade-off between flexibility and performances. Three factors combine to determine the characteristics of an FPGA: quality of its architecture, quality of the CAD tools used to map circuits into the FPGA and its electrical technology design. A first look at the FPGA hardware is provided in Chap. 3. The chapter provides in-depth analysis of programmable logic components and interconnection blocks in FPGA design and how they are interconnected to function as a generic reconfigurable system. This chapter establishes the basic understanding that FPGAs are semiconductor devices and contain programmable logic components connected by a regular, hierarchical programmable interconnect system. The distinguishing characteristic of FPGAs is their on-field programmability, which allows the logic functionality of an FPGA to be re-programmed even after the manufacturing process. FPGAs are used for rapid prototyping of digital circuits. The design and test of digital systems are very time-efficient and cost-effective with FPGAs. It also discusses about the logic components in the FPGA, mostly consists of memory elements such as registers or even complete blocks of memory that can be configured to hold any desired state. As we know, FPGAs were used mostly for prototyping and emulation systems in the design process of digital system design and ASICs. However, recently, FPGAs have become popular for a variety of mainstream products in networking, telecommunication, digital signal processing and in consumer electronics. FPGAs can be classified based on the technology using to program it.

FPGA architectures have been intensely investigated over the past two decades. A major aspect of FPGA architecture research is the development of Computer Aided Design (CAD) tools for design and implementation of fast and high density FPGAs and mapping applications to it. It is well established that the quality of an FPGA-based implementation is largely determined by the effectiveness of accompanying suite of CAD tools. Benefits of an otherwise well-designed, feature-rich FPGA architecture might be impaired if the CAD tools cannot take advantage of the features that the modern FPGA design provides. Thus, CAD algorithm research is essential to the necessary architectural advancement to narrow the performance gaps between FPGAs and other computational devices like ASICs.

Two-dimensional CAD flow of mesh-based and tree-based FPGA architectures are described and analyzed in Chap. 4. This chapter provides a perfect starting point for FPGA designs to evaluate and understand the algorithms and data-structures used in designing the software for FPGA placement and routing.

Chapter 5 provides the study of the existing variants of 2D tree-based FPGA architecture and the impact of 3D migration on its topology. We have seen numerous studies showing the characteristics of tree-based interconnect networks, how they scale in terms of area and performance and empirically how they relate to particular designs. Nevertheless, we have not had any breakthrough in optimizing these network topologies to exploit the advantages in area and power consumption and neither know how to deal with the larger wire-length issues that impede performance of tree-based FPGA architecture. Through the course of this book, we try to make the readers understand that, it is nearly impossible to optimize the area and speed, unless we break the very backbone of the tree-based interconnect network and resurrect again by using 3D technology. The 3D-ICs can alleviate interconnect delay issues by offering flexibility in system design, placement and routing. A new set of 3D FPGA architecture exploration tools and technologies developed to validate the advance in performance and area. Modern FPGAs have become a viable alternative to cell-based design technology by providing reconfigurable computing platforms with improved performance and higher density. While the reconfigurability provides flexibility, two-dimensional FPGAs also lead to area and performance overhead in comparison to cell-based custom integrated circuits (ICs). Thus, to combine the advantages of both FPGAs and custom ICs, modern FPGAs have emerged as an attractive solution for system-on-chip implementations. Modern FPGAs include design components such as digital signal processors, on chip memory blocks, multipliers, adders and entire processors. In Chap. 6 our primary focus is on validation of architecture exploration and optimization methodologies of 3D homogeneous and heterogeneous tree-based and mesh-based FPGAs.

A 3D-IC system consists of disparate materials with considerably different thermal properties including semiconductor, metal, dielectric and possibly polymer layers used for inter-plane bonding. Although the power consumption of these circuits is expected to decrease due to the considerably shorter interconnects, the power density increases since there is a greater number of devices per unit volume compared to a 2D circuit. As the power density increases, the temperature of the planes non-adjacent to the heat sink of the package can rise, resulting in degraded performance or thermal gradients that can accelerate wear out mechanisms. Design methodologies at various stages of the IC design flow, such as synthesis, floorplanning, placement and routing, which maintain the temperature of a circuit within specified limits or alleviate thermal gradients among the planes of the 3D circuit, are therefore necessary. Two key elements are required to establish a successful 3D thermal management strategy: a 3D thermal model, to characterize the thermal behaviour of a circuit and design techniques that alleviate thermal gradients among the physical planes of a 3D-IC system, while maintaining the operating temperature within acceptable levels. The primary requirements of a thermal model are high accuracy, low complexity and reasonably fast, while thermal design techniques should produce high-quality circuits without incurring long computational design time. To reduce the complexity of the modelling process, standard methods to analyse heat transfer, such as finite difference, finite element and boundary element methods, have been adopted to evaluate the temperature of a 3D circuit. Simpler analytic expressions have also been developed to characterize the temperature within a 3D system. The discussion culminates in Chap. 7 where design and implementation three-dimensional thermal model and thermal design techniques to improve the thermal profile and 3D-IC system and Chap. 7 focus its attention more on thermal analysis of 3D FPGAs.

Design techniques for three-dimensional (3D) ICs considerably lag the significant strides achieved in 3D manufacturing technologies. Advanced design methodologies for two-dimensional (2D) circuits are not sufficient to manage the added complexity caused by the third dimension. Consequently, design methodologies that efficiently handle the added complexity and inherent heterogeneity of 3D circuits are necessary. These 3D design methodologies should support robust and reliable 3D circuits while considering different forms of vertical integration, such as system-in-package and 3D-ICs with fine grain vertical interconnections. Global signalling issues, such as clock and power distribution networks, are further exacerbated in vertical integration due to the limited number of package pins, the distance of these pins from other planes within the 3D system and the impedance characteristics of the through silicon vias (TSVs). In addition to these dedicated networks, global signalling techniques that incorporate the diverse traits of complex 3D systems are required. One possible approach, potentially significantly reducing the complexity of interconnect issues in 3D circuits, is by optimizing the number of vertical interconnects (TSVs). Design methodologies that exploit the diversity of 3D structures to further enhance the performance of multi-plane integrated systems are necessary. Chapter 8 introduce new 3D physical design methodology and verification tools developed. This chapter also discuss the development of 3D physical design methodology and tools using existing 2D CAD tools for the implementation of 3D tree-based FPGA demonstrator. During the course of design process, we addressed many specific issues that 3D designers will encounter dealing with tools that are not specifically designed to meet their needs. In contrast, the thermal performance is expected to worsen with the use of 3D integration. In this Chapter, we examined precisely how thermal behaviour scales in 3D integration and determine how the temperature can be controlled using thermal design techniques.

A concreted effort has been made to present three-dimensional integration and high performance tree-based FPGA design using newly developed 3D physical design tools and VLSI design methodologies. Three-dimensional integration is an interdisciplinary field that relies on many experts working together at every design level. Emphasis is placed on illustrating the interaction among the different field. For example, 3D thermalware physical design described in Chap. 7 is a classic case of thermal, mechanical and electrical engineers working together to develop high performance three-dimensional integrated circuits. Few emerging research areas and possible future lines of research and applications of 3D-IC described in Chap. 9.

Dubai, United Arab Emirates
March 2015

Vinod Pangracious

Acknowledgments

I would like to thank all members of the Electrical and Computer Engineering Department at American University in Dubai and Laboratoire d'Informatique de Paris 6 (LIP6) at University of Pierre et Marie Curie Paris VI, for their continued support in writing this book.

Contents

Acronyms

3DIC	Three-Dimensional Integrated Circuits
ASIC	Application Specific Integrated Circuit
BFT	Butterfly-Fat-Tree
CMOS	Complementry Metal Oxide Semiconductor Field Effect Transistor
DRC	Design Rule Check
EDA	Electronic Design Automation
FPGA	Field Programmable Gate Array
GaAs	Gallium Arsenide
HFPGA	Hirarchical Field Programmable Gate Array
ITRS	International Technology Roadmap for Semiconductors
KGD	Known Good Dies
LVS	Layout Versus Schematic
MCNC	Microelectronics Center of North Carolina
MM	More Moore
MOSFET	Metal Oxide Semiconductor Field Effect Transistor
MtM	More-than-Moore
NRE	Non-Recurring Engineering
RF	Radio Frequency
SiP	System in Package
SiGe	Silicon-Germanium
SIA	Semiconductor Industries Association
SoC	System on Chip
SOI	Silicon on Insulator
SSI	Silicon–Silicon Interconnect
TSV	Through Silicon Via
VHDL	Very High Speed Hardware Description Language
VLSI	Very Large Scale Integration

Chapter 1
An Overview of Three-Dimensional Integration and FPGAs

Abstract The capabilities of many digital electronic devices are strongly linked to Moore's law: processing speed, memory and functional capacity and even the number and size of pixels in digital cameras. All of these are improving at roughly exponential rates as well. This exponential improvement has dramatically enhanced the impact of digital electronics in nearly every segment of the semiconductor industry, and is a driving force of technological and social change in the late 20th and early 21st centuries. This chapter discusses the historical evolution of semiconductor industry from 2D CMOS based technologies to today's three-dimensional (3D) integrated circuits using 3D vertical interconnects. Our main focus in this book is to explain the need and the development of tools and technologies that supports the utilization this emerging technology to improve the performance and manufacturability of high density Field Programmable Gate Arrays (FPGAs).

1.1 Introduction

The capabilities of many digital electronic devices are strongly linked to Moore's law: processing speed, memory and functional capacity and even the number and size of pixels in digital cameras. All of these are improving at roughly exponential rates as well. This exponential improvement has dramatically enhanced the impact of digital electronics in nearly every segment of the semiconductor industry, and is a driving force of technological and social change in the late 20th and early 21st centuries. Moore's Law is named after Intel co-founder Gordon E. Moore, who described the trend in his 1965 paper [1]. In it, Moore noted that the number of transistors in integrated circuits had doubled every year from the invention of the integrated circuit in 1958 until 1965 and predicted that the trend would continue *for at least 10 years*. Moore's prediction has proven to be uncannily accurate, in part because the law is now used in the semiconductor industry to guide long-term planning and to set targets for research and development. Historically, CMOS scaling has provided the means to realize higher performance with every technology node, as predicted by Moore's law. Ever since the 90 nm node, the gate length of MOSFETs (Metal-Oxide-Semiconductor-Field-Effect-Transistors) has entered the nano regime. The

V. Pangracious et al., *Three-Dimensional Design Methodologies for Tree-based FPGA Architecture*, Lecture Notes in Electrical Engineering 350, DOI 10.1007/978-3-319-19174-4_1

45 nm technology has become the mainstream since 2008, and 22 nm technology with Tri-gate (FinFET) transistors in 2012 and 14 and 10 nm with similar transistor technology expected in 2015 and 2016 respectively.

In 1998, the SIA (Semiconductor Industries Association) was joined by corresponding industry associations in Europe, Japan, Korea, and Taiwan to participate in a 1998 update of the Roadmap and to begin work toward the first International Technology Roadmap for Semiconductors (ITRS), published in 1999. The overall objective of the ITRS is to present industry-wide consensus on the *best current estimate* of the industry's research and development needs out to a 15-year horizon. For more than half a century these scaling trends continued, and expected it to continue until at least 2020. However, the 2010 update to the ITRS has growth slowing at the end of 2015, after which time transistor counts and densities are to double only every 3 years. Accordingly, since 2007 the ITRS has addressed the concept of functional diversification under the title *More than Moore* (MtM). This concept addresses an emerging category of devices that incorporate functionalities that do not necessarily scale according to Moore's Law, but provide additional value to the end customer in different ways. The MtM approach typically allows for the non-digital functionalities e.g., *RF communication, power control, passive components, sensors, actuators* to migrate from the system board-level into a particular package-level *SiP* or chip-level *SoC* system solution. It is also hoped that by the end of this decade, it will be possible to augment the technology of constructing integrated circuits (CMOS) by introducing new devices that will realize some *beyond CMOS* capabilities. However, since these new devices may not totally replace CMOS functionality, it is anticipated that either chip-level or package-level integration with CMOS may be implemented.

1.1.1 More Moore (MM)

The International Technology Roadmap for Semiconductors has emphasized in its early editions the *miniaturization* and its associated benefits in terms of performances and the traditional parameters in Moores Law. This trend for increased performances will continue, while performance can always be traded against power depending on the individual application, sustained by the incorporation into devices of new materials, and the application of new transistor concepts. This direction for further progress is labeled *More Moore or MM*. The multitude of new integration technologies opens many new possibilities for building an integrated electronic systems in a confined space and with high efficiency in terms of power dissipation and performance. In particular, the use of the third dimension in the backend/package allows combining products from different semiconductor as well as MEMS technologies. Thus, these advanced integration technologies link the requirements for high performance (More-Moore technologies or MM) with the demand for functional and technological diversity (More-than-Moore technologies or MtM). As we look at the years 2020–2025, we can see that the physical dimensions of CMOS manufacture are expected to be crossing below the 10 nm threshold. It is expected that as CMOS device dimensions approach the 5–7 nm range it will be difficult to operate any transistor structure that is utilizing the metal-oxide semiconductor (MOS) physics as the basic principle of operation. Of course, we expect that new devices, like the very promising semiconductor tunnel transistors, will allow a smooth transition from traditional CMOS to this new class of devices to reach these new levels of high performance ultra scale device integration. However, it is becoming clear that fundamental geometrical limits will be reached in the above timeframe. By fully utilizing the vertical dimension, it will be possible to stack layers of transistors on top of each other, and this 3-Dimensional (3D) approach will continue to increase the number of components per square millimeter even when horizontal physical dimensions will no longer be amenable to any further reduction. It seems important, then, that we ask ourselves a fundamental question: *How will we be able to increase the computation and memory capacity when the device physical limits will be reached?* It becomes necessary to re-examine how we can get more information in a finite amount of space.

1.1.2 More Than Moore (MtM)

During the blazing progress propelled by Moore's Law of semiconductor logic and memory products, many complementary technologies have progressed as well, although not necessarily scaling to Moore's Law. Heterogeneous integration of multiple technologies has generated *added value* to devices with multiple applications, beyond the traditional semiconductor logic and memory products that had lead the semiconductor industry from the mid 60s to the 90s. A variety of wireless devices contain typical examples of this confluence of technologies, e.g. logic and memory devices, display technology, micro-electrico-mechanical systems (MEMS), RF and Analog/Mixed-signal technologies (RF/AMS), etc. It should be emphasized that *More-than-Moore or MtM* technologies do not constitute an alternative or even a competitor to the digital trend as described by Moores Law. In fact, it is the heterogeneous integration of digital and non-digital functionalities into compact systems that will be the key driver for a wide variety of application fields. Whereas MM may be viewed as the brain of an intelligent compact system, MtM refers to its capabilities to interact with the outside world and the users as illustrated in Fig. 1.1.

In recent years, however, several bottlenecks have appeared as we have continued to scale down to sub- nm technology nodes and the question is, if the traditional technology scaling method alone will be able to overcome the performance and cost issues of the future IC manufacturing caused by interconnect delay and latency issues. The ITRS roadmap predicts 3D integration as a key technology to solve this so-called wiring crisis [2]. Several semiconductor industries and research institutes have demonstrated 3D integration process [3, 4]. Even though there are still no commercial true 3D-IC application in the market, it has become apparent that there is a strong demand for such future application such as memories, processors and logic devices. In addition to enabling of the further improvement of transistor integration densities (*More-Moore*), 3D integration is a well accepted platform for

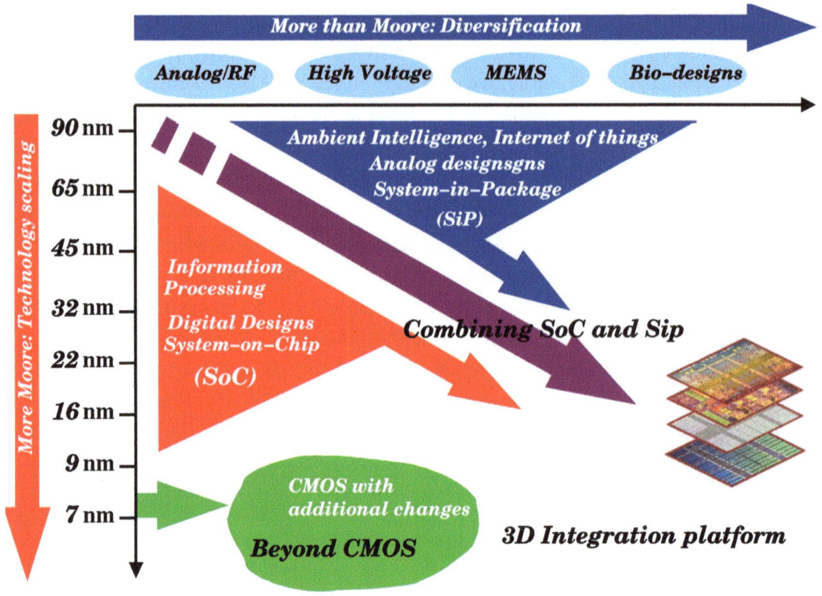

Fig. 1.1 The need for integrating digital and non-digital functionalities in an integrated system is translated as a dual trend in the International Technology Roadmap for Semiconductors: miniaturization of the digital functions (More Moore) and functional diversification (More-than-Moore), 3D integration represent a convergence of SoC and SiP disciplines [2]

More than Moore applications with their essential need for integration of heterogeneous technologies. Three-dimensional integration technology increases the number of active layers and optimizes the interconnect network vertically. The main advantage of 3D-IC technology is that it significantly enhances interconnect resources and increases logic density. If used correctly, 3D-ICs provides improved bandwidth and throughput, as well as reduced wire length. For N_{layers} stacking, in the best scenario, if the inter-layer vias are ignored, average wire length would be expected to drop by a factor of $(N_{layers})^{1/2}$. Both wire resistance and wire (RC) delay would drop by a factor of (N_{layers}). It also allow integration of dissimilar materials, process technologies, and functions onto one platform.

Our main focus in this book is to explain the development of tools and technologies that supports the utilization this emerging technology to improve the performance and manufacturability of high density Field Programmable Gate Arrays (FPGAs). FPGA chips offer an attractive solution for improving the design productivity through re-use of the same silicon implementation for a wide rage of applications. FPGA is programmable and can be reconfigured for yield improvement and defect tolerance. These features become absolutely necessary when CMOS technology scales down to nanometer scale, because the yield of the fabrication of semiconductor components hardly ever reach 100 %. FPGA consist of configurable logic blocks and I/O blocks that are interconnected by a configurable routing network. FPGA is configured to implement circuits by writing into the configuration memory that are embedded throughout the FPGA and defines the logical function of each block and connections within the configurable routing resources. Reconfigurability of FPGAs is fundamentally different from traditional general-purpose microprocessors. Microprocessors are attractive for their flexibility. An Application Specific Integrated Circuit (ASIC) is a device that is customized to a specific application. Since the exact nature of the application is known beforehand, ASIC hardware resources are designed to provide the highest performance implementation for the application. The price paid by ASICs because of their superlative performance characteristics is flexibility. Once an ASIC has been manufactured, it is impossible to modify it to implement another application, different from the one it was intended for. Further, since the Non-Recurring Engineering (NRE) costs involved in designing and manufacturing an ASIC are comparatively high, it is generally not feasible to design and fabricate ASICs in low volumes. Since their introduction in the mid eighties, FPGAs evolved from a simple, low-capacity gate array technology to devices [5, 6] that provide a mix of coarse-grained data path units, microprocessor cores, on chip A/D conversion, and gate counts by millions. Today, FPGAs become important actors in the computational devices domain that was originally dominated by microprocessors and ASICs. Just like microprocessors, FPGA-based systems can be reprogrammed on a per-application basis. At the same time, FPGAs offer significant performance benefits over microprocessor implementations for a number of applications. Although these benefits are still generally an order of magnitude less than equivalent ASIC implementations, the low NRE costs, fast time-to-market, and flexibility of FPGAs make them an attractive choice for low-to-medium volume applications.

1.2 Technological Initiatives and Contribution

FPGAs are consistently improving in capacity and performance, and are now among the most popular devices in the market. With their regular structure, they also scale easily to future technologies. However, FPGAs are still facing serious challenges in terms of delay, power consumption, and logic density compared to ASICs. FPGA is estimated to be over ten times less efficient in logic density, over three times worse in delay, and over three times higher in power consumption compared to a functionally equivalent ASIC [7–11]. Despite of their design cost advantage, FPGAs impose large area overhead when compared custom integrated silicon alternatives (ASICs). To illustrate the magnitude of this problem, we refer to the work presented in [11] where authors measure the gap between FPGAs and ASICs in terms of logic density, circuit speed and power consumption. The major performance and power bottleneck of the FPGA is the programmable interconnects and routing elements inside FPGA, which have been found to account for up to 80 % [9] of the total delay and up to 85 % [12] of the total power consumption and consume almost 90 % [13] of total silicon area, when both local and global interconnects are considered.

There is considerable demand for high performance FPGAs with low power consumption and area. One promising way to improve FPGA performance, logic density and power consumption is to incorporate three-dimensional (3D) integration, which increases the number of active layers and optimizes the interconnect network delay using vertical interconnects. There are few research initiatives for the design and implementation of 3D Mesh-based FPGAs [14–16]. For Mesh-based 3D FPGA, every layer in a 3D chip implements a normal 2D FPGA and this type of stacking reduces the average Manhattan distance between logic blocks, which leads to shorter interconnect resources. Consequently, 3D integration method is an attractive technology to improve the performance and density of FPGAs. Other gains, such as reduced design footprint and the ability to integrate different technologies, further favor 3D FPGAs. Used correctly, 3D integrated circuits provides improved bandwidth and throughput by reducing interconnect wire-length. However Mesh-based FPGA has a planer island style architecture which suites very well for a two-dimensional (2D) FPGA implementation. The major gains reported from the research and experimental demonstrations of 3D Mesh-based FPGA are not yet reached the scale of advantages and improvements expected according to ITRS roadmap [2], since the overall FPGA area and power consumption increases in 3D architecture, nevertheless the delay is reported to have reduced by 38 % [16]. Figure 1.2 presents the complexities involved in design and manufacturing of high density 3D chips. The true 3D implementation should bring holistic improvement in all areas of chip development starting from design to manufacturing. Many of the early designs and demonstrators of 3D FPGAs did not show much improvement in area and power consumption. In this book we try to revisit the traditional Tree-based FPGA architecture and main

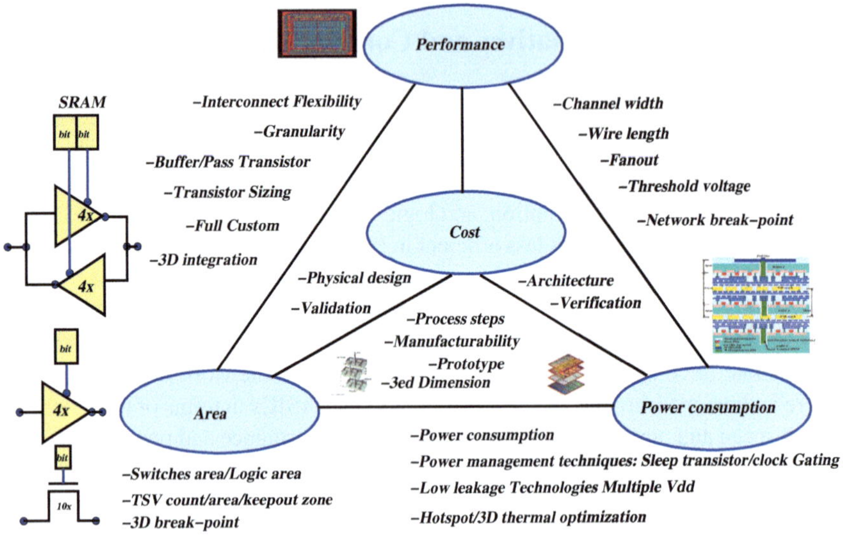

Fig. 1.2 FPGA Design and implementation challenges: The future is in the 3rd dimension

stream industrial FPGA architectures to conduct a feasibility study using 3D technology to improve logic density, area, speed and power consumption. The Tree-based multilevel interconnect architecture is one of traditional routing architecture of FPGA and multiprocessor system on chip (MPSoC) based systems. However it is not implemented in any of the industrial FPGA or MPSoC systems due to the large wire delays associated with the interconnect impede the performance of the system.

1.2.1 *Modified Tree-Based Interconnect*

The aim of this book is to revisit the traditional and industrial FPGA architectures to propose a suitable interconnect architecture model to design and manufacture 3D FPGAs with improved logic density and speed. An efficient *butterfly-fat-tree* interconnection network structure is proposed in [17–20] to design and implement high density FPGAs. A detailed analysis of area and switch requirements of Mesh- and Tree-based FPGA architectures presented in [19, 20]. The reported results shows that the 2D Tree-based architecture improve total area by 56 % and reduce the total switch requirement by 59 % compared to 2D Mesh-based FPGA architecture. Nevertheless the wire delay increases logarithmically as the Tree grows to higher level and this makes the 2D physical design implementation of Tree-based FPGA architecture a daunting task. The complexities associated with the development 2D Tree-based architecture layout is presented in [18]. In this book, we propose new design solutions and exploration methods using 3D technology to improve logic density, area, and power consumption of 3D FPGAs using Tree-based multilevel interconnect architecture. The main sections of 3D FPGA design presented in this book as follows.

1.2.2 *Tree-Based Interconnect Partitioning*

Interconnect network partitioning is the best way to reduce the length of interconnects and theirby improveing speed and power consumption. Two independent network partitioning methodologies are proposed to design and implement 3D Tree-based FPGA.

- *Vertical partitioning:* the programmable interconnect network is partitioned vertically by placing the *break-point* at the highest level ℓ_v of the Tree-based programmable interconnect network to balance the silicon area and power consumption across multiple tiers of the 3D chip
- *Horizontal partitioning:* the main objective is to optimize the critical path delay and improve logic density. The horizontal *break-point* is placed at a particular tree level ℓ_h based on the design and manufacturing constraints to achieve interconnect delay optimization using TSVs.

The location of the level ℓ_v is always fixed at highest tree level, however the location of level ℓ_h is decided based on the architecture and wire delay requirements.

1.2.3 3D FPGA Design and Implementation Methodology

To design and implement 3D multi-tier Tree-based FPGA, we developed a set of 3D physical design methodology and tools using Global Foundries 130 nm technology node modified to use Tezzaron's TSV technology [21, 22]. The design flow covers all areas of 3D design, including the design partitioning, merging multiple tiers (gds files) and design sign-off analysis. In addition, we also address the specific issues that 3D designers will encounter dealing with tools that are not specifically developed to meet their needs. We developed additional design support programs to enable the designer to perform 3D DRC/LVS and TSV implementation using six metal back-end offline (BEOL) technology.

1.2.3.1 3D FPGA Physical Design Tools

This book describe the development of an automated 3D physical design methodology including a VHDL code generator based on Tree-based FPGA architecture description and design constraints. The VHDL code generator is based on a hierarchical design approach that partitions the design into smaller sections, which implement clusters separately and assemble them together at the final design phase. The physical design is performed using Global Foundries 130 nm technology node (Tezzaron 3D Design platform). A timing evaluation system based on Mentor's circuit simulator *Eldo* is attached to design module to accurately estate the networks delays.

1.2.3.2 3D FPGA Architecture Exploration Tools and Methodologies

Our goal through the development of this book is to develop an efficient placement and detailed routing tool for 3D Tree-based FPGAs. Using this tool, we investigate the impact of 3D integration on delay, area and power consumption, in addition to wire-length reduction because wire-length alone cannot be relied on as a metric for 3D integration benefits. The main features of the architecture exploration tool is mentioned below.

- Feasibility study of different network partitioning methods to find suitable interconnect architecture for 3D staking.
- 3D Tree-based FPGA architecture optimization tool: This tool is developed as an add-on facility to 3D place and rout tool to find minimum interconnect and TSV requirements for the implementation of 3D FPGA. A Rent's Rule [23, 24] based wire-length distribution model is used to design the architecture optimization tool.

- FPGAs are not really FPGAs any more instead, they are arrays of programmable gates plus DSP slices, ALUs and transceivers etc. The 3D Tree-based exploration is augmented to study 3D Heterogeneous Tree-based FPGAs as well. The tools have capabilities to analyze the the placement and location of *hard-blocks* to optimize the speed and area of the 3D FPGA chip.

1.2.4 Unified Mesh of Tree Architecture

Recently we have witnessed 2.5D and 3D FPGA product demonstrations from leading FPGA research institutions and manufacturing industries. These new FPGA architecture also introduces many opportunities and challenges to meet with the expectations of increasing functionality of modern FPGA chip designs. In this book we propose a variant of Tree-based FPGA architecture with qualities of both Mesh and Tree-based interconnect architectures. Our previous studies [19, 20] shows Tree-based FPGA has better logic density and area advantage compared to Mesh-based FPGAs. In this study, we examine the possibility of unifying the advantages of modified Tree- and Mesh-based interconnect architecture into one platform to improve density, area, and speed of 3D FPGAs.

1.2.4.1 Architecture Improvement, Tools and Methodologies

In this book we propose a 3D interconnect network implementation based on a modified Mesh-of-Trees (MoT) topology for FPGA architecture design as an extension of Tree-based FPGA architrecture. We further optimized the MoT-based interconnect architecture using *long wire segments* with adjustable *span* to transform it into a viable architecture for the design and implementation of high density 2.5D multi-FPGA and 3D stacked multi-tier FPGA based systems. Exploration and physical design tool flows developed to demonstrate the performance improvement and area advantage of 2.5D and 3D MoT-based FPGA architecture. The two possible variants of MoT-based FPGA architecture implemented are as follows.

- **2.5D Multi-FPGA using Mesh of Tree Architecture**: We developed exploration and validation tools to explore the impact of *% of wires cut* and *no of cuts* on performance and area of 2.5D multi-FPGA based systems.based of MoT-based FPGA architecture. Using the 2.5D tools flow, we can demonstrate the improvement in area and performance of 2.5D multi-FPGA with 1–3 cuts and different variants of interposer-based inter-FPGA connections.
- **3D Multi-tier FPGA using Mesh of Tree Architecture**: A 3D architecture exploration tool (place and route) developed to estimate the area and delay reduction in 3D stacked MoT-based FPGA architecture. We also implemented an MoT-based architecture optimization tool using Rent's Rule to find the optimal architecture to estimate the impact of *% of long wires* on channel width W of the 3D MoT-

based FPGA architecture. Unlike the other 3D Mesh-based FPGA architectures, in 3D MoT-based FPGA architecture, we have a direct relation between vertical interconnects and channel width. This relationship is established to optimize the horizontal and verticals routing resource requirements. This book will not discuss the deatils of the MoT-Based FPGAs, since the focus of this book is on 3D implementation of Tree-based FPGA architectures.

1.3 Book Organization

A brief overview of the contents of the book as follows. This Chapter provides a brief introduction about the main fcous theme, thrust area and contributions made to it for the development of high density 3D FPGAs. Chapter 2 starts with brief introduction to 3D integration and discusses the main challenges and opportunities of 3D technology in areas like process integration and CAD tools development. This chapter also discusses few new practical solutions to solve technological and CAD level issues in 3D physical designs process. The main purpose of this work is evaluate the interconnect architecture of traditional and mainstream FPGA architectures and to propose an alternative 3D interconnect architecture or suggest the required modification to design and manufacture high density FPGAs using 3D Technology. Chapter 3 discusses the pros and cons of different FPGA architectures and propose new architectural changes to the existing FPGA architectures. An interesting state of the art of CAD tools for the exploration of 2D Mesh- and Tree-based FPGA architectures presented in Chap. 4. This chapter also report the importance of research and development of CAD algorithms for high density FPGA development. Chapter 5 presents the state of the art 3D FPGA. Many new ideas and implementations regarding 3D-ICs and FPGAs shows positive developments across the world to migrate the present day technology to the third dimension. This chapter also present the 3D design and architecture exploration methodology developed for the implementation of 3D Tree-based FPGAs.

The Chap. 6 provides a detailed analysis of the improvement in speed, area and optimization of 3D Tree-based FPGA architecture. Once an architecture is verified for its performance, its also curious to know how this architecture is going to behave when the few architectural parameters changes. This chapter also provides answers to those questions about the impact of LUT size and cluster size on performance of 3D FPGA chip. FPGAs have evolved dramatically over the past 10 years, as they have taken advantages of new process technologies and architectural innovations. One particular issues has arisen due to the lack of a robust architectural exploration tool that can model heterogeneous *hard-blocks* such us memories and multipliers. Chapter 6 describes a detailed description of validation exploration tools developed of 3D Tree-based heterogeneous FPGA architectures and also provides the result analysis of critical path delay and architecture optimization.

The power consumption of 3D-ICs is expected to decrease due to the interconnect length reduction. However the power density increases since the distance between the devices decreases per unit volume as compared to a 2D layout. Consequently, the temperature also rises. Thermal aware design or hardware design techniques should be implements at various stages of the 3D-IC design flow, such as synthesis, floorplanning, placement and routing to maintain the temperature of the chip with acceptable limits. The Chap. 7 provides detailed description and analysis of different thermal aware design techniques and hardware-based methods developed to improve thermal profile of 3D Tree-based FPGAs. The development of 2D physical design for Tree-based FPGA interconnect is a daunting task. Chapter 8 sheds light into those issues designers face and also describe how to resolve them using 3D technology. This chapter provides two interesting network partitioning methodologies for Tree-based interconnect to mitigate the traditional long wire-length issues associated Tree-based interconnect architectures. Three-dimensional design and technology is famous for it ability to improve speed, power consumption and silicon footprint of semiconductor chips. This chapter introduce new 3D design and varification tools and methodologies of 3D FPGA design and implementation. The Chap. 9 narrates the summary of the thesis and provides future lines of research work.

References

1. G. Moore, Cramming more components onto integrated circuits. Proc. IEEE **86**(2), 82–85 (1998)
2. SIA: Semiconductor Industries Association, *The International Technology Roadmap for Semiconductor* (SEMATECH, Austin, TX, 2009)
3. P. Garrou, C. Bower, P. Ramm, *Handbook of 3D Integration* (Wiley-VCH, 2008). ISBN: 978-3-527-32034-9
4. V.F. Pavlidis, E.G Friedman, *Three-Dimensional Integrated Circuit Design* (Morgen Kaufmann, 2009). ISBN: 978-0-12-374343-5
5. Altera, *Stratix V device overview* (2013), www.altera.com
6. Xilinx Inc, *Two flows for partial reconfiguration: module based or difference based* (2004), http://www.xilinx.com/bvdocs/appnotes/xapp290.pdf
7. J. Rose, R. Francis, D. Lewis, P. Chow, Architecture of field-programmable gate arrays: the effect of logic functionality on area efficiency. IEEE JSSC **25**(5), 1217–1225 (1990)
8. V. Betz, J. Rose, How much logic should go in an FPGA logic block?. IEEE Des. Test Comput. **15**(1), 10–15 (1998)
9. E. Ahmed, J. Rose, The Effect of LUT and cluster size on deep-submicron FPGA performance and density. IEEE Trans. Very Large Scale Integr. VLSI Syst. **22**(3), 288–298 (2004)
10. M. Lin, A.E. Gamal, Y.-C. Lu, S. Wong, Performance benefits of monolithically stacked 3D FPGA, in *Proceedings of the 2006 ACM/SIGDA 14th International Symposium on Field Programmable Gate Arrays*, Monterey, California, USA, 22–24 Feb 2006, pp. 113–122
11. I. Kuon, J. Rose, Measuring the gap between FPGAs and ASICs. IEEE Trans. Comput. Aided Des. Integr. Circuits Syst. **26**(2), 203–215 (2007), http://dx.doi.org/10.1109/TCAD.2006.884574 (IEEE Council on Electronic Design Automation)
12. F. Li, D. Chen, L. He, J. Cong, Architecture evaluation for power-efficient FPGAs, in *Proceedings of the ACM/SIGDA International Symposium on Field Programmable Gate Array*, Nov 2003, pp. 175–184
13. A. DeHon, Reconfigurable Architectures for General-Purpose Computing, Ph.D. dissertation, Department of Elect Engg and Computer Science, Massachusetts Institute of Technology, 1996

14. C. Ababei P. Maidee and K. Bazargan, Exploring potential benefits of 3D FPGA integration, in *Field Programmable Logic and Application*, vol. 3203 (Springer, Berlin, Germany, 2004), pp. 874–880
15. K. Siozios, A. Bartzas, D. Soudris. Architecture level exploration of alternative schmes targeting 3D FPGAs: a software supported methodology. Int. J. Reconfig. Comput. **2008** (2008)
16. K. Siozios, V.F. Pavlidis, D. Soudris, A Novel framework for exploring 3-D FPGAs with heterogeneous interconnect fabric. ACM Trans. Reconfig. Technol. Syst. **5**(1) (2012)
17. A. DeHon, Unifying mesh- and tree-based programmable interconnect. IEEE Trans. Very Large Scale Integr. VLSI Syst. **12**(10), 1051–1065 (2004)
18. A. DeHon, R. Rubin, Design of FPGA interconnect for multilevel metallization. IEEE Trans. Very Large Scale Integr. VLSI Syst. **12**(10), 1038–1050 (2004)
19. Z. Marrakchi, H. Mrabet, C. Masson, H. Mehrez, Mesh of tree: unifying mesh and MFPGA for better device performances, in *NOCS-2007*, pp. 243–252 (2007)
20. Z. Marrakchi, H. Mrabet, U. Farooq, H. Mehrez, FPGA interconnect topologies exploration. Int. J. Reconfig. Comput. **2009** (2009)
21. S. Gupta, M. Hilbert, S. Hong, R. Patti, *Techniques for Producing 3D ICs with High-Density Interconnect* (Tezzaron Semiconductor, Naperville, IL, 2005)
22. R. Patti, Advances in 3D memory and logic devices, in *IMAPS International Conference on Device Packaging* (TAI3, Scottsdale, AZ, March 2010)
23. B. Landman, R. Russo, On a pin versus block relationship for partitions of logic graphs. IEEE Trans. Comput. **20**(12), 1469–1479 (1971)
24. J. Pistorius, M. Hutton, Placement rent exponent calculation methods, temporal behaviour and FPGA architecture evaluation, in *Proceedings of the International Workshop on System Level Interconnect Prediction* (Monterey, Calif, USA, April 2003), pp. 31–38

Chapter 2
Three-Dimensional Integration: A More Than Moore Technology

Abstract Three-dimensional integrated circuits (3D-ICs), which contain multiple layers of active devices, have the potential to dramatically enhance chip performance, functionality, and device packing density. They also provide for microchip architecture and may facilitate the integration of heterogeneous materials, devices, and signals and offer a promising solution for reducing both silicon footprint and interconnect length without shrinking the transistors. However, before these advantages can be realized, key technology and CAD challenges of 3D-ICs must be addressed. More specifically, the process required to build circuits with multiple layers of active devices and CAD tools used for design and validation of such circuits. Several such methodologies and CAD tools associated with the design fabrication of 3-D ICs are discussed in this chapter. Few successful 3D-IC design methods and CAD tools and benefits of applying 3D design to the future reconfigurable systems are also discussed in this chapter.

2.1 Introduction

The ongoing demand for greater functionality resulting in multiple IC products, longer off-chip interconnects ravage the performance of microelectronic systems. The advent of System-on-Chip (SoC) in the mid 1990s primarily addressed the increasing delay of the off-chip interconnects. Integrating all of the components on a monolithic substrate enhances the overall speed of the system, while decreasing the power consumption. To assimilate disparate technologies, however several difficulties must be surmounted to achieve high yield for the entire system. Additional system requirements for the radio frequency (RF) circuitry, passive elements, and discrete components, such us decoupling capacitors, which are not easily integrated due to performance degradation or size limitations. While Moore's law [1] and the pursuit of ever increasing transistor counts is well known in IC design and manufacturing circles, what is seldom brought to light for others, are the escalating cost and technology challenges associated with this pursuit. Smaller transistors and larger dies have been reasonable answer to this quest in the past. Stacked dies using wire bond connections and flip-chips have even been employed to create system-in-package (SiP) solutions

V. Pangracious et al., *Three-Dimensional Design Methodologies for Tree-based FPGA Architecture*, Lecture Notes in Electrical Engineering 350, DOI 10.1007/978-3-319-19174-4_2

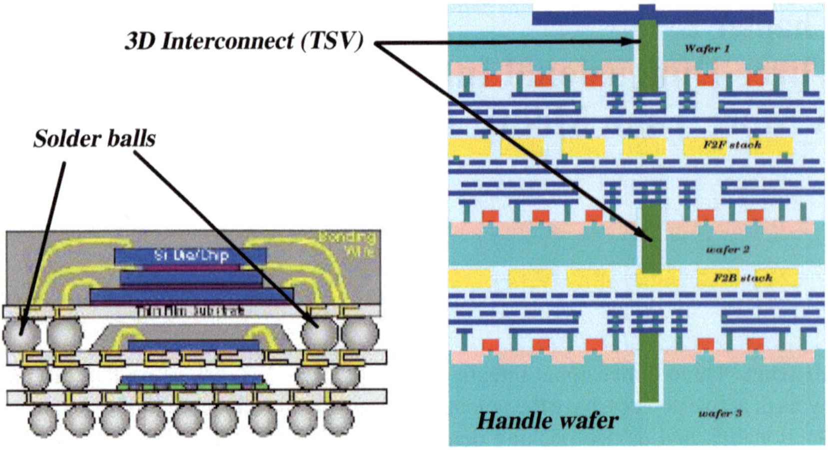

Fig. 2.1 Interconnects bond wires in typical System-in-package (SiP) and 3D-IC

that meet the needs of some. Looking for alternative solutions for next generation designs, that meet the performance, integration, form-factor, manufacturability, and cost requirements, may have begun to look at going up rather than out. With this trend, the Three-dimensional (3D) integration using through-silicon via (TSV) technology has gained much attention. Once the domain of specialist applications, more mainstream users, such as memories, microprocessors ans specialized logic designs are now being considered as TSV candidates. The advantages of 3D-IC integration are better electrical performance, low power consumption, lower area and weight and high performance (Fig. 2.1).

2.1.1 Opportunities for Three-Dimensional Integration

Performance requirements such as increased bandwidth, reduced latency, and lower power consumption are driving the adoptions of 3D-IC designs. A complete 3D-IC implementation is usually envisioned as a stack of active chips using TSVs to connect through each chip down to a package substrate. TSV designs represent a convergence of SoC and SiP disciplines, providing designers the means to significantly increases the bandwidth between the logic chip and the memory especially with wide memory interfaces that cannot be achieved with bond wires, as well as the ability to mix and match dies that not only use different process node, but also different manufacturing technologies such as SiGe, SOI, CMOS low voltage, CMOS high voltage, Biploar, GaAs, etc. The ability to combine different dies in a single stack enable to acquire needed functionality to provide high-quality, proven die. What is new in 3D-ICs is the ability to place vertical interconnections (TSVs) in a dense array, without the strict perimeter constraints imposed by an equivalent wire-bonded design. Utilizing stacked chips, particularly in memory-intensive designs, allows designers to stay at today's reasonable process nodes for each die and derive the benefit or proven volume manufacturing processes.

Three-dimensional (3D) die stacks and high-bandwidth silicon packaging technology using emerging through silicon vias (TSVs), thinned silicon, and fine-pitch silicon-silicon interconnections (SSIs) make use of a wide variety of technology structures, materials, and processes. Universities, consortia, and industry have driven research and early demonstrations for a decade. TSV and SSI interconnection density can scale in excess of six orders of magnitude, making the technology widely applicable from simple to very complex applications. At academic research institutes and semiconductor industries, new 3D test-vehicle (i.e., demonstrator) designs followed by manufacturing, assembly, and characterization studies continue to provide technologists with an understanding of structure and process-integration capabilities and limitations. Results from these technology studies provide guidance on 3D design rules, structures, processes, tests, and reliability, which can support the manufacturing of 3D products and provide data that we may use to determine technology directions. Practical technology fabrication and integration approaches need to consider targeted TSV and SSI interconnection density, silicon thickness, and power densities. In addition, decisions with respect to options such as TSV conductor material, SSI integration material, and use of die-on-die, die-on-wafer, or wafer-to wafer process approaches need to be made with regard to interconnection redundancy, die size, yield, cost, and test methodology.

The inherent advantage of 3D integration is the drastic decrease in interconnect length, particularly the long global interconnects, which directly results in increased speed [2–6]. We can understand this by simple geometry analysis for 3D-ICs. A given squre area A has maximum Manhattan wirelength id $2\sqrt{A}$. The same area is split into two tiers reduces the wirelength to $\sqrt{2}\sqrt{A} + l_v$ where l_v is the length of via between tiers. In general, n layers gives a maximum Manhattan wirelength of $2\sqrt{\frac{A}{n}} + (n-1)l_v$. Figure 2.2 illustrate the graphical representation of wire-length reduction which the original 2D chip implemented using 3D technology with n tiers. The interconnect power is also reduced as the capacitance of the wires decreases

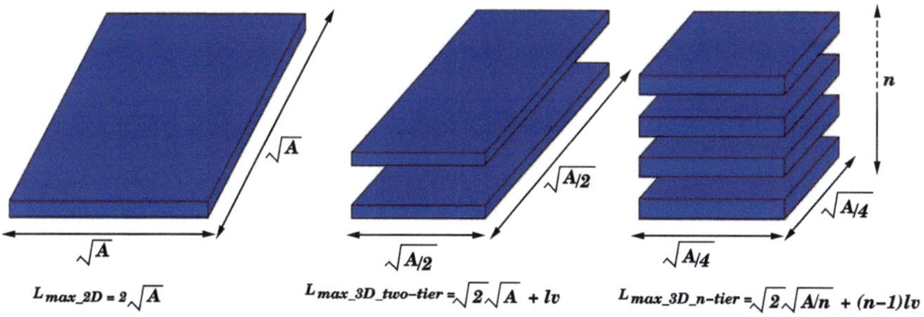

Fig. 2.2 Illustration of wire-length reduction where the original 2D chip implemented using 3D technology

[7, 8]. Additionally, the total power dissipated by an interconnect system is further decreased as the number of repeaters inserted along the interconnect is reduced [9]. Finally, coupling among intraplane adjacent interconnects is lower due to decreased length, improving signal integrity. The 3D-IC based systems provides the capability to include disparate technologies [10], greatly extending the capabilities of modern systems-on-chip (SoC). This defining feature of 3D-ICs offers unique opportunities for highly heterogeneous and sophisticated systems [11, 12]. A vast pool of applications such as medical, wireless communications, military, and low-cost consumer products, exists for vertical integration, as the proximity of the system components caused by the third dimension is suitable for either the high performance or low power ends of the SoC application space [13]. This heterogeneity, however, greatly complicates the interconnect design process within a multi-tier system, as potential design methodologies need to manage the diverse interconnect impedance characteristics and process variations caused by the different fabrication processes and technologies employed in the different physical tiers.

Three-dimensional circuits can be conceptualized as the bonding of multiple wafers or bare dice. The distinctive difference between an SiP and a 3D IC is the granularity of the vertical interconnects. Different bonding styles between the planes within a 3D system are also possible Face-to-Face (F2F), Face-to-Back (F2B), and Back-to-Back (B2B) [14, 15]. Examples of SiP structures and various bonding styles for 3D circuits are illustrated in Fig. 2.1. Each of these bonding styles is likely to include through silicon vias (or interplane vias) with different physical dimensions. Consequently, the density of the vertical interconnects can vary not only among different 3D circuits but also among the physical planes within a 3D circuit.

2.2 Historical Evolution of 3D System Integration

The proposal of doubling the number of transistors on an IC chip every 24 months by Gordon Moore in 1965 (Moores law) [1] has been the most powerful driver for the development of the microelectronics industry in the past 45 years. This law emphasizes lithography scaling and 2D integration of all functions on a single chip, perhaps through system-on-chip (SoC) as schematically shown in the left-hand side of Fig. 2.3. On the other hand, the integration all these functions can be achieved through 3D-IC integration [13, 16–18] as illustrated in Fig. 2.4.

Through Silicon Via (TSV) is the heart 3D-IC integration [19]. Though the 1956 Nobel Laureate in Physics, William Shockley invented TSVs more than 50 years ago in U.S. Patent #3,044,909, filed in 1958 and issued in 1962, but it was not intended for 3D-IC integration and it took half a century for the production technology to reach the level of expertise that would actually permit making TSVs. From a die with hundreds of transistors in the 1960s to dies approaching billions of circuits in 2014, on-chip integration has continued to require lithographic advancements for circuit and increase in wire density has led to increasing the number of wiring levels on the chip. Over decades of semiconductor scaling, on-chip integration has far

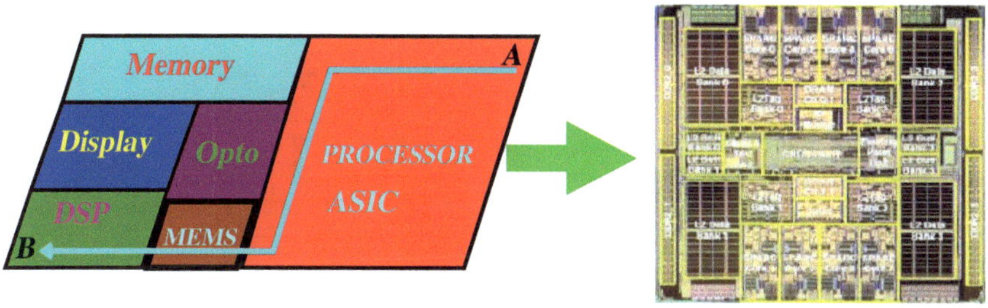

Fig. 2.3 Typical 2D System-on-Chip (SoC) integration

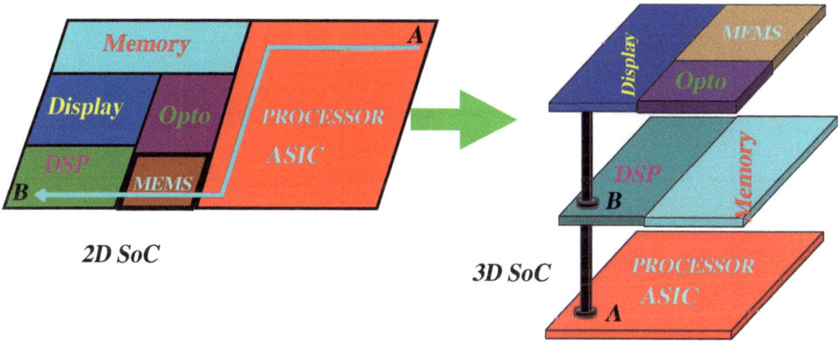

Fig. 2.4 The difference in wire length between 2D Soc and 3D SoC, the wire length between blocks A and C in 2D SoC and 3D SoC

out-stripped off-chip 3D integration and growth in I/O interconnections. For off-chip interconnections over the last five decades, I/O interconnections grew from tens of I/O interconnections to about several thousands of I/O interconnections for the most complex die manufactured today [20–22]. Figure 2.5 shows the evolution of 3D-IC technology along with the resulting relative I/O interconnection density for 3D design and implementation. The emerging 3D integration approaches can be implemented different packaging form factors to combine TSVs, thinned silicon and interposer technology as required to achieve higher interconnection density. High bandwidths may be achieved using 3D chip integration, 3D die stacking, or 3D silicon packaging in which each form factor offers high interconnection density (10^4/cm^2 to 10^8/cm^2). Therefore, trade-offs between best system form factors will be dependent on factors such as system architecture, manufacturing costs, and test and assembly integration yields.

Looking toward the future, industry and academic researchers are developing wafer-to-wafer and die-to wafer stacking techniques for the fabrication of devices that leverage the z-direction but eliminate the need for multiple packages [23]. Additionally, these techniques reduce interconnect delays, form factors, and power consumption while allowing integration of numerous heterogeneous devices. In the wafer-to-wafer approach, circuitry is divided into sections that are built onto separate

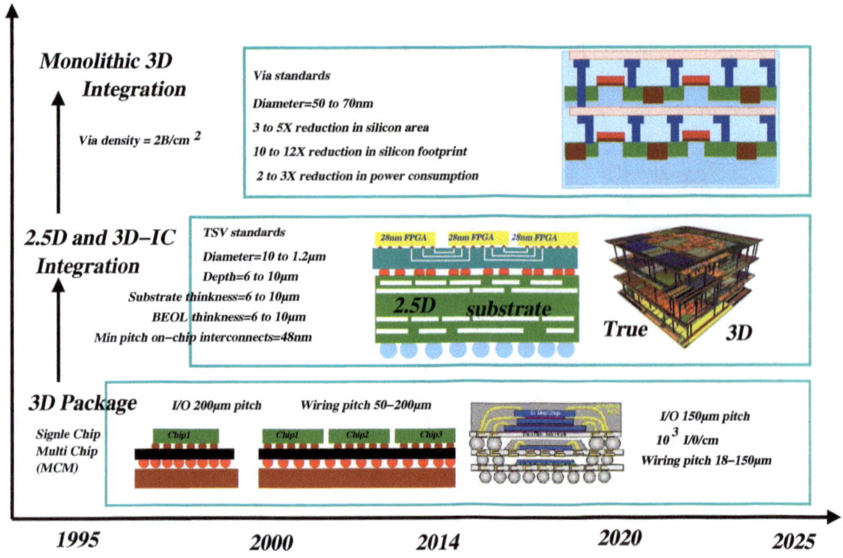

Fig. 2.5 Evolution of 3D integration and *vertical* interconnect technology

wafers using standard processing methods. The wafers are then post-processed for through-silicon interconnections (TSVs), creating the vertical connectors. The wafers are aligned, bonded, thinned, and diced into individual devices. In October 2006, several equipment manufacturers, led by Alcatel, EV Group, Semitool, and XSil, formed a consortium, dubbed EMC-3D, to address the technical and cost issues associated with the creation of TSV interconnect technology for die stacking and wafer-to-wafer attach. In the die-to-wafer variation, a known good die (KGD) is bonded to a wafer. This approach is preferred in configurations that require three or more dies in a stack. Privately held Ziptronix Inc, a spin-out business of the Research Triangle Institute, advanced the state of the art in the die-to-wafer methodology when it introduced, late in 2005, a direct bond interconnect technology (a covalent room-temperature bond) that replaces through-die vias (TSVs), increases electrical connection density, and reduces interconnect delays.

Research investigations have explored a wide variety of structures, processes, and bonding approaches. Researchers recognize the importance of

- developing fine-pitch vertical interconnections using TSVs,
- developing thinning technology for silicon and interconnection technology that joins thinned silicon dies into die stacks and that joins dies to silicon packages,
- developing wafer-to-wafer bonding technologies.

In addition to power delivery and signal interconnections, investigators have also included approaches for thermal cooling and modeling of heat removal from thinned silicon structures and fine-pitch interconnections. Vertical interconnect (TSV) technology is a key focus area and an enabler for the evolution of 3D-IC design and packaging. As already indicated, TSV technology is one of the key interconnect solutions enabling a vertical method of electrical connectivity for various 3D-IC configurations such as stacked die and wafer-level packaging. In TSV investigations, technical reports have included studies in which researchers sought submicron TSV diameters for compatibility with wafer front-end-of-line (FEOL) and back-end-of-line (BEOL) wafer fabrication or alternatively for silicon-based package solutions. TSV diameters and pitches have ranged from large sizes, such as about 10–100 μm via diameter and silicon thickness of about 50–300 μm, down to via diameters of less than 1–10 μm with corresponding silicon thicknesses ranging from about 50 μm down to about 6 μm silicon thicknesses. Reported TSV conductors have included tungsten, copper, composite, paste, doped polysilicon, as well as other electrical conductors. For example, [24] gave a TSV technical presentation on 10 μm copper conductors utilizing TSVs for electrical interconnection ata 20 μm pitch.

Fine-pitch interconnection, also at a 20 μm pitch for silicon-on-silicon connections with TSVs, has also been reported by [25] and also variety of bonding and electrical interconnection approaches between silicon die in thinned silicon die, die stacks, or packages using silicon reported in [26, 27]. In these interconnection examples, anisotropic conductive polymers were used to bond 25 μm thinned dies with 50 μm pitch AuSn bumps. Technical publications have also reported fine-pitch solder connections to copper as a means either to stack thinned silicon chips to other silicon dies or to join dies to silicon packages [25–28]. An application that leverages TSVs and fine-pitch interconnections with demonstration of functioning memory die stacks has also been presented [29]. The main future challenge for TSV technology relates to its ability to maintain performance parameters, such as signal integrity or heat management, as data rates climb. However, a number of companies have been able to demonstrate efficient TSV electrical interconnect solutions that meet data rates on the order of 10 Gb/s. Many technology suppliers as ZyCube, Intel, Samsung and IBM are currently optimizing the manufacturability and reliability of their 3D-IC fabrication process. Tezzaron's *Super-Via* technology, initially a post backend-off-line process with Cu–Cu bonding [16] was abandoned due to failures of the used copper TSVs (5 μm diameter). In consequence they changed their process now to tungsten filled *Super-Contacts* with 1.2 μm diameter [30, 31].

2.3 Vertical Interconnect Technology Development (TSV)

Through Silicon Via or TSVs are a critical enabler for both wafer-to-wafer and die-to-die stacking for which low-inductance, high bandwidth vertical interconnects are needed in silicon. Applications may require only a few, thousands, or millions of vertical interconnections, a number that is very product dependent and is affected by architecture, desired product specifications, silicon thickness, materials, structures, and processes. the range in size includes diameters from less than 1.2–90 μm. The silicon thickness ranges from less than 6 μm to a full wafer thickness of 730 μm, with

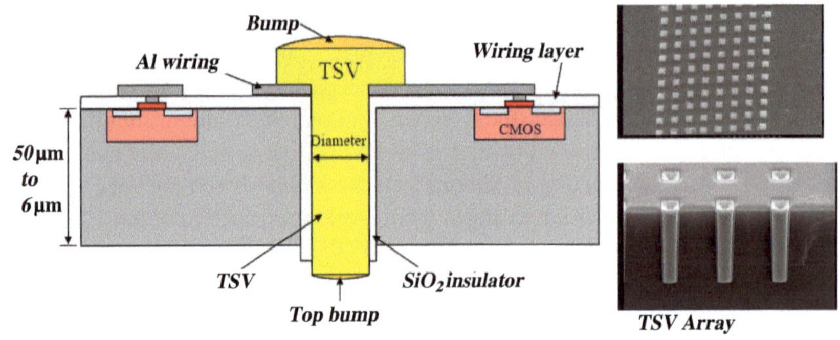

Fig. 2.6 Representation of Through Silicon Via (TSV)

most studies having been performed with 150 μm thicknesses or less. Material evaluations have included copper, tungsten, and composite materials. Figure 2.6 shows examples of TSV cross-sections.

As illustrated in Fig. 2.6, a through-silicon via (TSV) is a vertical via that completely passes through a silicon die. Its main purpose is to establish electrical connectivity between devices in two different dies in a 3D-IC stack. There is presently no consensus on the most efficient bonding technique which largely depends on the application requirements [30, 32, 33]. There are various bonding techniques [34] which range from direct oxide bonding, metal to metal (Cu–Cu) bonding, with different variants and adhesives. Still, the most prevalent technique to stack TSV based dies is a micro-ball based bonding. Micro-balls or micro-bumps are the most appropriate for present 3D applications since the density of TSVs is not very high (IK–10K/chip), their locations are predetermined and micro-bump based stacking is presently more reliable that alternative techniques. Depending on when the TSVs are fabricated, two major types of TSV exist: via-first and via-last, as illustrated in Fig. 2.7.

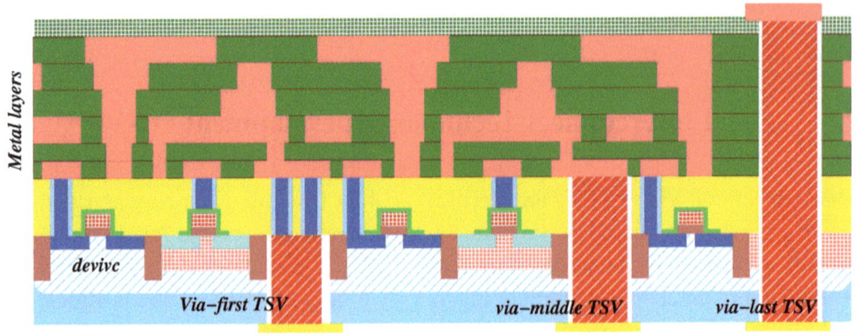

Fig. 2.7 An illustration of Via-first, via-mid and via-last TSV technology process

- Via-first: TSV's are fabricated before CMOS or Si frontend (FEOL, Front-End-Of-Line) device fabrication processing.
- Via-middle: TSV's are fabricated after the Si frontend (FEOL) device fabrication processing but before the backend (BEOL-back-end-of-line), interconnect process.
- Via-last: TSVs, are fabricated after or in the middle of the Si Backend (BEOL) or bonding, essentially when the wafer is finished.

The dimensions of via-first TSVs are typically smaller (1–10 μm diameter), with aspect ratios (=height:diameter) of 3:1–10:1. A key benefit to the via first approach is that companies using it don't need to worry about spoiling expensive wafers at the R&D stage because they can use bare Si or SOI wafers. For Si interposer wafer development related to heterogeneous stacking, via-first is still being developed and used. In this case of via-last TSVs, the processing can be done at the foundry or packaging house and there is the possibility to start TSV processing from the top surface of the wafer (Front-side processing) where the active transistor layouts are placed. The via-last TSV diameter is wider (10–50 μm), with aspect ratios of 3:1–15:1. There are two main technologies for *drilling* TSVs: dry etching or Bosch etching, and laser drilling. Polysilicon, copper, and tungsten are the most popular materials for TSV fill. Silicon dioxide is a popular material for the liner that sits between the TSV and silicon substrate for insulation purpose. From the perspective of physical design, via-first TSVs are less intrusive because they interfere only with the device, M1, and top layers, whereas via-last TSVs interfere with all layers in the die as illustrated in Fig. 2.7. Via-first TSVs have their landing pads on M1 and the top metal layers, whereas via-last TSVs have their landing pads only on the top metal layers. These landing pads include keep-out-zone uniformly located around them to reduce coupling effects. The connection between via-first TSVs are made using local interconnect and vias in between adjacent dies, whereas via-last TSVs are stacked on top of each other as illustrated in Fig. 2.7. Therefore, via-first TSVs are usually used for signal and clock delivery, whereas power delivery network utilize via-last TSVs in general.

2.4 3D Integration: Manufacturing Methods

Various 3D integration technologies currently pursued by semiconductor industry and research institutions. There are many different integration and manufacturing schemes for 3D interconnects. One way to categorize the different integration schemes is by the orientation of the individual dies to each other. Figure 2.8 shows face-to-back (F2B) and face-to-face (F2F) integration. F2F integration does not require TSVs in general, however TSV can be used for I/O connections, whereas TSVs are required for F2B integration. For two-layer chip stacks, both integration schemes have some advantages and disadvantages. F2B configuration uses standard process for test, assembly and packaging, however F2F do not have a standard process. For multi-layer stacks, F2B has the advantage that after each bonding step

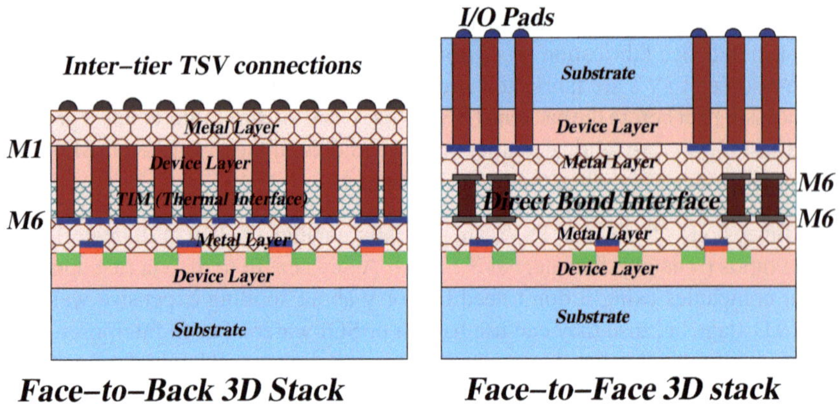

Fig. 2.8 3D stacking methods: F2F and F2B stacking configuration

the top device layer is face up so that the stacking unit process can be repeated multiple times. However, for F2B integration, the die/wafer has to have the final thickness already during stacking, as subsequent thinning is not possible. In F2F stacking configuration wafer thinning is not a requirement.

Another more popular way to categorize 3D integration schemes is based on the point at which the TSV is created during the manufacturing process. In the past, the only distinction was whether the via was manufactured before or after wafer thinning: via-first or via-last. Today, it is common to further distinguish whether the via was created prior to front-end processing, i.e., via first, or after front-end processing (but before wafer thinning), i.e., via-middle. Figure 2.9 shows a typical process flow for via-middle manufacturing. Another important distinction within the various integration schemes is based on wafer or die level processing: chip-to-chip (C2C), chip-to-wafer (C2W) and wafer-to-wafer (W2W). C2C has mainly been used for high performance, high margin devices. For lower margin devices like consumer electronics, C2C is not very suitable due to single die processing. Of course, W2W integration allows wafer-level processing after stacking. W2W integration gives the highest throughput and the highest alignment accuracy. But W2W integration requires that the dies have the exact same size, and it has the inherent risk that a defective die is bonded to a good die, thereby destroying the whole stack. C2W is a hybrid process and combines the single die placement with the feasibility of wafer-level processing after die placement. With C2W integration, it is possible to stack dies of different sizes. With a modular design and chip architecture, it must be assumed that dies typically will have different sizes. For heterogeneous integration in particular, C2W is the method of choice as currently only silicon devices are manufactured on 300 mm wafers, while all other semiconductor materials are being manufactured on smaller wafer sizes. In addition C2W enables testing of every die prior to stacking, which allows true *known good die* manufacturing. Figure 2.10 shows the difference between C2W and W2W integration. A fourth stacking method

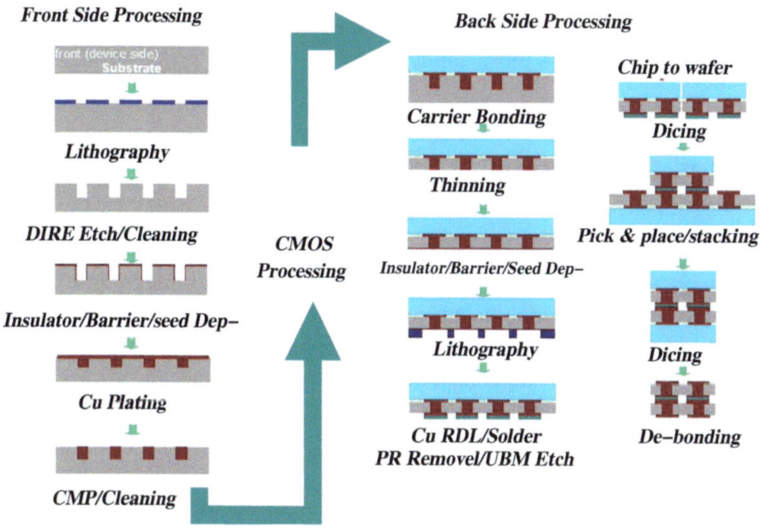

Fig. 2.9 A typical 3D Via-mid stacking process integration: *Front-side* and *Back-side* processing methods

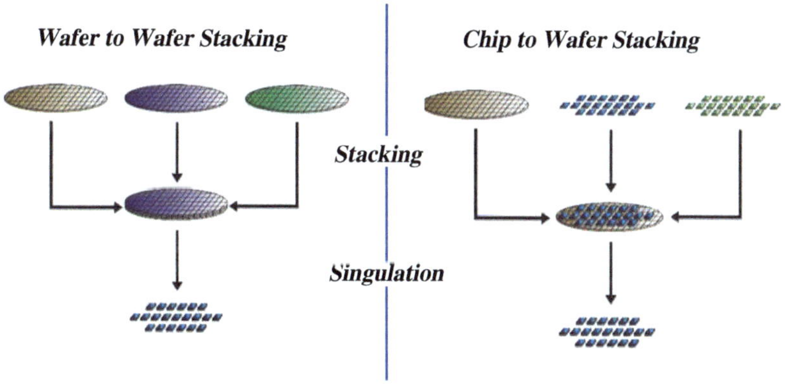

Fig. 2.10 3D Integration methods: Wafer-to-Chip and Wafer-to-Wafer stacking

and well advanced in today's FPGA industry is silicon interposer-based integration (called horizontal or 2.5D integration), vertical die to die stacking (also called 3D stacking), and a range of mixed configurations. An interposer-based stacking gained popularity last year because of several attractive applications [35] as well as its technological feasibility. It might not be sufficiently effective for other applications, however, such as the memory on logic [36] or the logic splitting application, where the logic is split between two or more dies that are then put on top of each other for shorter interconnections. Also memory can be split in such a way that read-write logic is on one chip while the cells are on the other. A true 3D stacking is needed for maximum performance in those applications.

2.5 Challenges in 3D Physical Design

The introduction of the third dimension has significantly increased the complexity of the integrated circuit design process. The 3D design and integration faces enormous challenges in both manufacturing technology and physical design. The major challenge is to define the characteristics of the verticals interconnects and the constraints that this type of interconnects poses on physical design process and other typical problems are reuse of existing 2D-IP blocks, testability, CAD tools and thermal issues. While thermal integrity is a critical issue in all high performance chip design, since the system reliability is strongly dependent on the temperature and this problem is even more significant for 3D designs due to the high power density in the stacked arrangement. Increasing the number of tiers that can be integrated into a single 3D system is a primary objective of 3D integration. A 3D system with high-density vertical interconnects is therefore indispensable. Vertical interconnects implemented as TSVs produce the highest interconnect bandwidth within a 3D system, as compared to wire bonding, peripheral vertical interconnects, and solder-ball arrays. Alternatively, the density of this type of interconnect dictates the granularity of the interconnected layers of the system, directly affecting the inter-tier communication bandwidth. Other important criteria should also be satisfied by the TSV fabrication process. A fabrication process for vertical interconnects should produce reliable and inexpensive TSVs. A high TSV aspect ratio, the ratio of the diameter of the top edge to the length of the via, may also be required for certain types of 3D circuits. The effect of forming the TSVs on the performance and reliability of neighboring active devices should also be negligible.

The electrical characteristics of the TSVs are of primary importance in 3D-ICs and are considerably different from the horizontal interconnect segments [37], as described by recent electrical models [38]. This situation is due to the structure of these interconnects and the diverse technologies, such as CMOS and SOI, that can exist in a 3D system. Producing low resistance and capacitance TSVs is a fundamental objective of manufacturing technologies. Finally, not properly characterizing the contribution of the TSVs to the delay of the critical inter-tier interconnect can result in significant inaccuracy in the performance of a 3D system [39]. Consequently, these structures must be carefully considered during the 3D physical design process. The thermal traits of the TSVs are also significant, as these vias can affect the thermal behavior of a 3D-IC. TSVs can be used to provide high thermal conductivity paths to facilitate the flow of heat from the upper tiers to the tiers attached to the heat sink, maintaining the temperature of a 3D circuit within acceptable levels. Materials with low thermal resistance, such as copper, are therefore preferred.

2.5.1 Complexity of 3D Physical Design Tools and Their Limitations

The solution space for classical physical design methodologies increases significantly in 3D systems, as the physical distance of two circuit cells is reduced not only by placing these cells near each other on the same tier but also by placing the cells in vertically adjacent locations. This situation results in a formidable increase in the number of solutions that can be explored, resulting in an exponential growth in the computational time. The increase in the number of metal layers yields similar computational issues for the routing task [40]. Computationally efficient heuristic algorithms are the primary tool to manage the dramatic increase in the solution space for 3D circuits. Methods such as simulated annealing (SA) and genetic algorithms complete the mosaic of the 3D physical design process [35, 41]. The main challenge is 3D physical design is dealing with tools that are not specifically designed to meet their needs. There are several works presented in the literature that describe various 3D system design options and physical design algorithms for 3D-ICs, but very few in the area of 3D design demonstration and methodology.

One quick solution to the lack of physical design tools for 3D-ICs is to build so-called *pseudo* 3D tools, which are based on straightforward extension of existing tools for 2D-ICs [30]. These pseudo 3D tools are able to handle simple 3D designs, wherein existing 2D designs are simply stacked and connected without any major design change. A good example of this is 3D stacking of processor and memory dice, where the only change required is to add TSVs in the layouts to deliver signal, power, and clock in vertical directions. This can be done by adding TSVs in the layout whitepace or by slightly modifying the layout to leave space for TSVs and wires. These TSVs are then treated as pseudo IO pads in the layout. In addition, traditional objectives, such as wire length and area, are insufficient for 3D circuits, particularly heterogeneous multi-tier integrated systems. Since these systems can combine disparate technologies, such as radio frequency (RF), analog, and digital circuits, other objectives, such as noise and signal integrity, need to be simultaneously considered in addition to conventional objectives. These objectives require the synergistic development of design methodologies, which previously were individually developed for each type of circuit. However, future trends in 3D-IC design calls for finer-grained 3D optimizations, such as 3D module floorplanning or 3D gate placement across the tiers for performance and power consumption improvement. In addition the number of tiers in the stack is expected to increase in order to meet the demand for higher-level system integration. These trend requires more powerful native 3D physical design tools that are built from ground up and are capable of handling many tiers and many TSVs simultaneously while addressing the current and emerging issues like cost, reliability, and manufacturability. In addition TSV and thermal-aware 3D design, verification, and analysis tools including 3D DRC/LVS, timing, power, signal integrity, power integrity and clock integrity analysis tools need to be seamlessly integrated and efficiently managed.

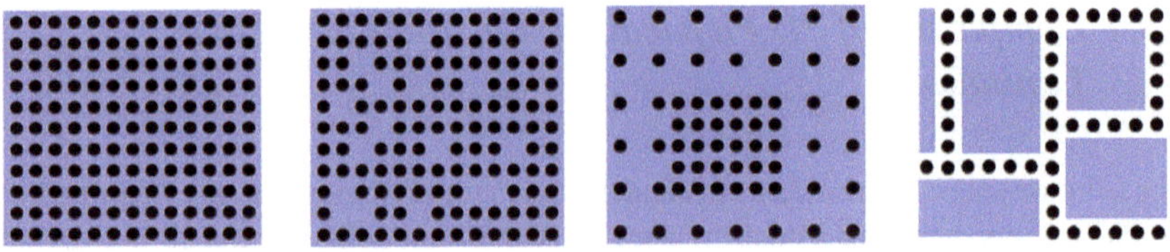

Fig. 2.11 TSV placement styles for 3D Stacked chips: Regular and non-regular placement for improving design quality and thermal profile of the chip

2.5.2 *TSV and Thermal Management*

TSV management is at the heart of physical design for 3D-ICs. Especially, the count and location of TSVs have significant impact on the quality and reliability of 3D-IC layouts. A recent study [42] shows that the overall wire length-up to a certain point reduces as more TSVs are used in the 3D layout. The number of TSVs used in 3D-IC layout entirely depends on how the design is partitioned into multiple dies. Research is needed to determine the optimal partitioning styles for given applications. Possible solutions include tradeoff studies among core-level, block-level, and gate-level partitioning across the dies in the 3D stack. Research is also required to investigate the impact of TSV location on 3D-IC design quality and reliability [43]. Possible solutions include trade-off studies between regular and non-regular TSV placement [43] with respect to these metrics, as illustrated in Fig. 2.11. TSV cost is another important factor that needs to be addressed during physical design. The cost of TSV depends on geometry-related data such as the pitch, diameter, and aspect ratio as well as the materials used for the TSV fill and liner. Moreover, the total number of TSVs used in the layout significantly affects the overall cost of the 3D-ICs. It is important to model and balance the trade-offs between the cost and other metrics such as performance, power, reliability, and manufacturability during physical design with TSVs.

A 3D system consists of disparate materials with considerably different thermal properties including semiconductor, metal, dielectric, and possibly polymer layers used for plane bonding. Although the power consumption of these circuits is expected to decrease due to the considerably shorter interconnects, the power density increases since there is a greater number of devices per unit volume as compared to a 2D circuit. As the power density increases, the temperature of the planes nonadjacent to the heat sink of the package can rise, resulting in degraded performance or thermal gradients that can accelerate wear out mechanisms [44, 45]. Design methodologies at various stages of the IC design flow, such as synthesis, floorplanning, and placement and routing, which maintain the temperature of a circuit within specified limits or alleviate thermal gradients among the tiers of the 3D circuit, are therefore necessary. Two key elements are required to establish a successful thermal management strategy: a thermal model, to characterize the thermal behavior of a circuit, and design techniques that alleviate thermal gradients among the physical layers of a 3D stack while maintaining the operating temperature within acceptable levels. The primary requirements of a thermal model are high accuracy and low complexity [46–48], while thermal design techniques should produce high-quality circuits without incurring long computational design time [49]. To reduce the complexity of the modeling process, standard methods to analyze heat transfer, such as finite difference, finite element, and boundary element methods, have been adopted to evaluate the temperature of a 3D circuit. Simpler analytic expressions have also been developed to characterize the temperature within a 3D system.

Thermal design techniques can be classified into two categories: thermal strategies that improve the thermal profile of a 3D circuit without requiring any redundant interconnect resources for thermal management and those methodologies that are an integral part of a more aggressive thermal policy that utilize thermal TSVs, sacrificing other design objective(s). The thermal aware design techniques uses the location and spatial distribution of TSV to accurately estimate the temperature profile of the 3D chip. As discussed earlier, TSVs are made of copper or tungsten and they are good thermal conductors. By carefully arranging the TSVs, it is possible to transfer the heat efficiently from the tiers far from heatsink towards the tiers placed near to heatsink. To do this, we need to include 3D thermal analysis tool to the physical design flows to create thermal aware design tools. As described in Fig. 2.11, depending of the type of design, a uniform or a non-uniform TSV distribution can be used to effectively transfer heat from one layer to another. There are also more aggressive thermal management methods using redundant interconnect resources. These TSVs are typically called thermal or dummy vias [10] to emphasize the objective of conveying heat rather than providing signal communication for circuits located on different physical layers. Thermal wires can also be employed to transfer heat [50]. Thermal wires correspond to those horizontal wires that connect regions with different thermal via densities through thermal inter-tier vias.

2.5.3 *Power and Clock Delivery in 3D-ICs*

On-chip power delivery is a major challenge in 3D-IC design. In 3D ICs, the on-chip power-ground (P/G) networks in several tiers are vertically connected with P/G TSVs, leading to much higher current demand per TSV. The number of TSVs used in the 3D P/G network is also limited so as to prevent placement and routing congestion. In addition, signal routing must be done carefully to prevent coupling noise between P/G TSVs and signal wires. This complex optimization problem usually results in larger area, more power consumption, and more noise, which leads to less performance and diminishing benefit of TSV-based 3D-IC technology. Research needed on P/G network synthesis, optimization, and analysis to addresses these issues while minimizing on-chip resource usage such as P/G wires, P/G TSVs. The sequential elements in 3D ICs (i.e., flip-flops and latches) are potentially located in all of the dies in the 3D stack. This poses a major challenge in delivering clock signal to all of them while reducing power consumption, skew, slew, and jitters. A recent study [51] shows that more clock TSV usage up to a certain point translates to more wire length reduction and thus power saving. However, clock TSVs, as in the case with signal and P/G (Power and Ground) TSVs, occupy layout space and causes coupling. Thus, clock TSV management becomes an important issue in 3D clock tree synthesis. In addition, the high thermal variations in 3D ICs induce a substantial amount of skew variation in the clock tree, which has adverse implications for the performance and reliability of 3D-ICs. The 3D clock tree itself is the longest wire in the circuit and contains many buffers to control skew and slew. Since the delay characteristics of clock wires, buffers, and TSVs are significantly affected by the temperature, care must be taken to ensure that the skew is kept minimum based on a given non-uniform thermal profile.

2.5.4 TSV-Induced Design for Manufacturability Issues

Primarily due to their large size compared with other layout objects, TSVs in 3D-IC layouts cause significantly non-uniform layout density distributions on the active, poly, and M1 layers. This density variation issue is expected to cause trouble during chemical-mechanical polishing (CMP) steps in the BEOL processing of the individual die, and requires new TSV-aware solutions. In addition, the printability of the devices and wires nearby TSVs will be affected in a non-negligible way. The CTE (coefficient of thermal expansion) mismatch between TSV copper and silicon causes significant stress to the devices nearby during manufacturing and operation of the 3D ICs. This in turn affects the timing characteristics of the devices and thus the overall circuit performance. The reliability of substrate and devices nearby TSVs is also affected because the thermal hotspots created in the regions cause repeated thermal expansion and contraction during 3D IC operation. This transient thermal behavior, together with the residual stress from TSV fabrication, may cause cracking and other physical damage in the substrate and devices. Research efforts required to address these issues during physical design. Possible solutions include TSV-aware CMP fill synthesis for the top and bottom metal layers, TSV stress-aware timing analysis and physical design [52], and TSV-aware substrate and device reliability modeling and optimization. TSVs are significantly larger than devices and local interconnects and thus complicates physical design and optimization for 3D layouts. Accurate electrical, mechanical, and thermal modeling of TSVs is essential in successful physical design of TSV-based 3D-ICs. In addition, full-chip layout construction and analysis for 3D-ICs should consider the impact of TSVs on performance, power, reliability, manufacturability, and cost.

2.5.5 Floorplanning for 3D Circuits

The predominant design objective for floorplanning a circuit has traditionally been to achieve the minimum area or, alternatively, the maximum packing density while interconnecting these blocks with minimum length wires. Most floorplanning algorithms can be classified as either slicing [53] or nonslicing [54, 55]. Floorplanning techniques belonging to both of these categories have been proposed for 3-D circuits [56–59]. An efficient floorplanning technique for 3-D circuits should adequately handle two important issues: representation of the third dimension and the related increase in the solution space. Conventional floorplanning assumes a single 2D layer on which several modules must be arranged. A wide verity of different algorithmic approaches have been used in order to solve the floorplanning problem. 3D floorplanning includes new 3D-specific characteristics that must be represented in the underlying data structures. For example, high output power modules need comprehensive consideration, such as thermal-driven floorplanning [60] and vertical dependencies arise in addition to horizontal ones.

There are two ways to represent the vertical dependencies. The first possibility is the multiple usage of classical data structures, so-called 2.5D methods. Here, additional mechanisms have to be implemented to consider vertical relations between module placed in different tiers, such as vertical alignments as well as overlapping and non-overlapping constraints. The representations include a discrete z-direction, such as the combined bucket and 2D array approach (CBA) in [61]. Vertical dependencies must incorporated directly into the data structure to prevents invalid solutions without time-consuming evaluations as well as to minimize the solution space. More recent data structures for floorplanning represent multilayer modules in the three dimensions. An example of such a data structure is 3D Slicing Tree described in [53, 62]. As illustrated in Fig. 2.12, different operations, such as module rotation and swapping, can be carried out efficiently to modify the given tree. A concatenation of these operations allows obtaining any possible slicing tree from any given slicing tree. However solutions from a 3D-Slicing Tree are limited to slicing floorplans.

2.5.6 Placement for 3D Circuits

Placement algorithms have traditionally targeted minimizing the area of a circuit and the interconnect length among the cells, while reserving space for routing the interconnect. In vertical 3D-IC integration, a placement dilemma arises in deciding whether two circuit cells sharing a large number of interconnects can be more closely placed within the same tier or placed on adjacent physical tiers, decreasing the interconnection length. Placing the circuit blocks on adjacent tiers can often produce a line with the shortest wire-length to connect these blocks. An exception is the case of small blocks within an SiP where the length of the inter-tier vias is greater than 100 μm [63, 64]. Placement methodologies have also been discussed where other objectives, such as thermal gradients among the physical tiers and the temperature of the tiers [65], are considered. Several approaches have been adopted for placing circuit cells within a volume [66–70]. Different types of circuit cells for various 3D technologies have been investigated in [36]. Layout algorithms for these cells have also been devised, demonstrating the benefits of 3D integration. Since TSVs consume

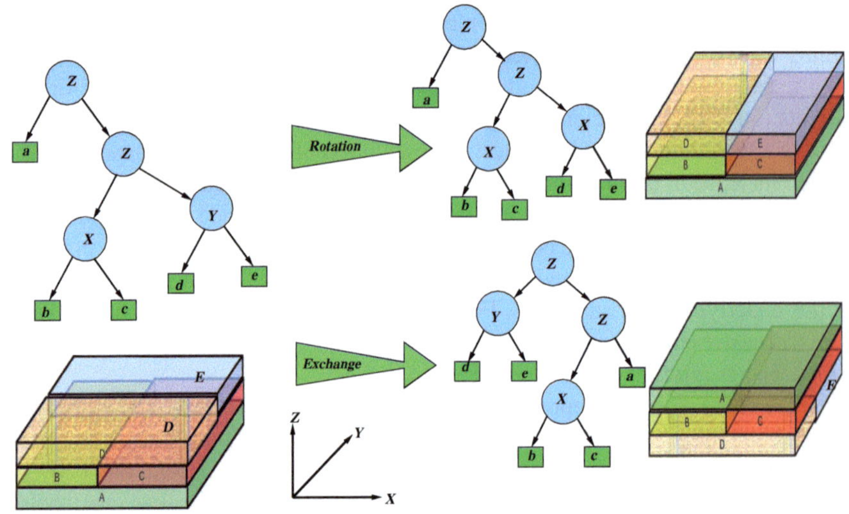

Fig. 2.12 Illustration of 3D Slicing Tree operation to permute a given 3 F floorplan: A rotation alters an inner node (representing a cut through the normal plain) resulting in a physical rotation of modules contained in the sub-trees of that node. An exchange swaps two sub-trees resulting in a physical exchange of modules contained in these sub-trees

silicon area, possibly increasing the length of some interconnects, an upper bound on this type of interconnect resource is necessary. Alternatively, sparse utilization of the vertical interconnects can result in insignificant savings in wire length. To consider the effect of the vertical interconnects, a weighting factor has been used to increase the distance in the vertical direction, controlling the decision as to where to insert the inter-tier vias [69]. This weight essentially behaves as a controlling parameter that favors the placement of highly interconnected cells within the same or adjacent physical tiers. Alternatively, TSVs are treated as circuit cells since these interconnects occupy silicon area [71] and are included in the individual cell placement process within each tier. Since this approach can result in two different locations for placing a TSV, as illustrated in Fig. 2.13, a weighted average distance between these two locations can be utilized to place a TSV [71]. Although these approaches consider the location of the TSV, the fundamental objective is to decrease the interconnect length. The maximum achievable reduction in the interconnect length for the longest on-chip interconnect is proportional to $\sqrt{n}$, where n is the number of tiers constituting a 3D circuit [8]. Any further improvement in the performance of the inter-tier interconnects can be obtained by considering the electrical characteristics of the TSV.

Multi-objective placement techniques for 3D circuits are necessary to generate efficient 3D floorplans. Additional objectives that affect both the cell placement and wire length are simultaneously considered. The force directed method is a well-known technique used for cell placement [72], where repulsive or attractive forces are placed on the cells as if these cells are connected through a system of springs. The force directed method has been extended to incorporate the thermal objective

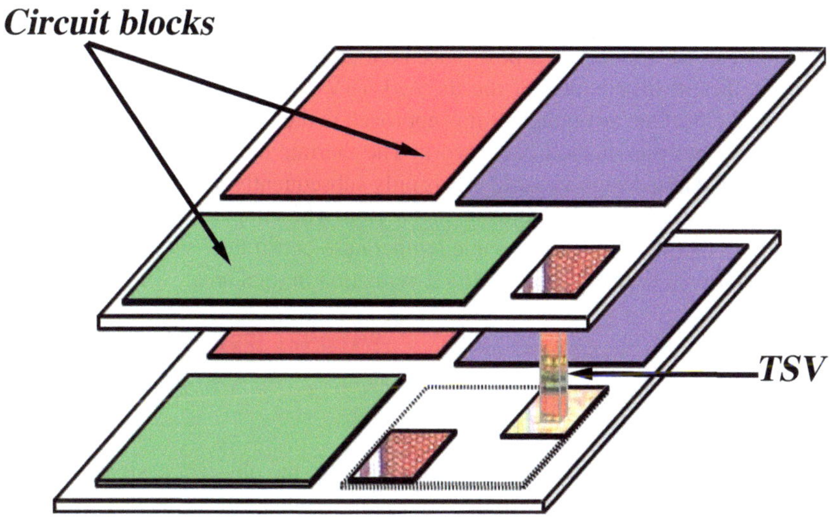

Fig. 2.13 Treating the TSVs as circuit cells on different planes can result in two different locations for placing a TSV. These locations define a region in which the TSV can be placed to satisfy different design objectives

during the placement process [73]. In this approach, repulsive forces are applied to those blocks that exhibit high temperatures (i.e., *hot blocks*) to ensure that the high-temperature blocks are placed at a greater distance from each other. The efficiency of this force directed placement technique has been evaluated on the MCNC [74] and IBM-PLACE benchmarks [75], demonstrating a 1.3 % decrease in the average temperature, a 12 % reduction in the maximum temperature, and a 17 % reduction in the average thermal gradient. The total wire length, however, increases by 5.5 %. As demonstrated by these results, this technique primarily achieves a uniform temperature distribution across each plane, resulting in a significant decrease in thermal gradients as well as the maximum temperature. The average temperature throughout a 3D-IC, however, is only slightly decreased. As a more practical example that demonstrates the need to include the thermal objective in 3D physical design techniques, consider an Intel Pentium 4 processor, which has been redesigned in two planes [76]. The increased power density due to stacking can increase the peak temperature within the 3D processor by approximately 26 °C, as compared to the original 2D system if thermal issues are ignored [76]. This increase can significantly degrade the performance and reliability of the processor. If the thermal objective is incorporated during the placement process, a negligible 2 °C increase is observed [76]. Alternatively, additional TSVs that do not function as a signal path can be utilized to further enhance the heat transfer process. The design objective is to identify those regions where thermal vias are most needed (hot-spots) and place thermal vias within those regions at the appropriate density. Such an assignment, however, is mainly restricted by two factors; the routing blockage caused by these vias and the size of the unoccupied regions or white space that exist within each tier. Although thermal via insertion can be applied as a post-placement step, integrated techniques produce a more efficient distribution of the thermal TSVs for the same temperature constraint [77, 78]. One advantage of this approach is the large granularity with which the thermal analysis method could work. The thermal conductivity of each region can be treated as a design variable that is only subsequently translated into a precise number of thermal vias placed inside this region. The integrated technique requires 16 % fewer thermal vias for the same temperature constraint, with a 21 % increase in computational time and an almost 3 % reduction in total area.

2.5.7 Routing for 3D Circuits

Routing is the most complex and least developed of the physical design techniques used in 3D circuits. During the routing stage, all terminals of the nets in the circuits netlist must be properly connected while respecting the constraints, such as design rules, routing resources capacities and optimizing routing objectives like minimize total wire-length, maximum timing slack etc. The multiple metal layers available for routing on each physical layers exacerbate the difficulty in routing a net connecting several circuit cells located on different layers. As these interconnects also compete with the transistors for silicon area, routing is a formidable task for 3D circuits. An early paper on routing 3D circuits demonstrated several issues related to this physical design task [79]. Consequently, several heuristics have been developed that address routing in the third dimension [80, 81]. The main difference between the 2D and 3D routing is caused by the multi-tier position of the net terminals that lead to net topologies which span more than one tier as illustrated in Fig. 2.14. This requires expensive inter-tier vias to be used in addition to regular signal vias which connect metal layers within the same tier. An effective approach for routing 3D circuits is to convert the routing inter-tier interconnect problem into a 2D channel routing task, as the 2D channel routing problem has been efficiently solved [82, 83]. A number of methods can be applied to transform the problem of routing the inter-tier vias into a 2D routing task, which requires utilizing a portion of the available routing resources for inter-tier routing (usually the top metal layers). Inter-tier interconnect routing can be implemented in five major stages including inter-tier channel definition, pseudoterminal allocation, inter-tier channel creation (channel alignment), detailed routing, and final channel alignment [80]. Additional stages route the 2D channels, both the inter-tier and intra-tier vias, and perform channel ordering to determine the wire routing order for the 2D channels.

Alternatively, multilevel algorithmic techniques [84] have been applied to route 3D circuits. The advantages of multilevel routing are the lower computational time and higher completion rates as compared to flat and hierarchical routers. Multilevel routing can be treated as a three stage process: a coarsening phase, an initial solution generation at the coarsest level (level p) of the grid, and a subsequent refinement process until the finest level of the grid is reached. Before the coarsening phase

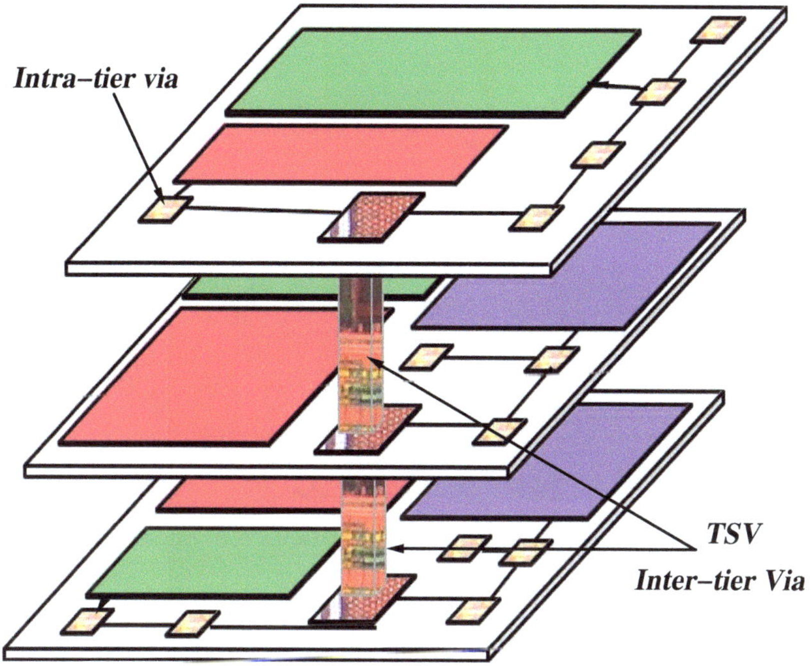

Fig. 2.14 Example route for a net in a three-tier 3D design

is initiated, the routing resources in each unit block of the grid are determined by a weighted area sum model. The routing resources are allocated during each coarsening step. The resources for the local nets within a block are transferred at each coarsening step. At the coarsest level, an initial routing tree is generated. This initial routing task commences with a minimum spanning tree for each multi-terminal net. A Steiner tree heuristic and a maze searching algorithm generate a 3D Steiner tree for each of these interconnects. Additionally, the TSVs are estimated for each block. During the last phase, the initial routing tree is refined until the finest level is reached. In this refinement phase, the signal (and thermal) TSVs are successively assigned and distributed within each block. The routing of the wires follows the refinement of the TSVs. At the finest level, a detailed router completes the routing of the circuit [84].

Multilevel routing for 3D-ICs has been extended to include the thermal objective [85, 86]. A thermal-driven 3D router using a multilevel routing approach composed of a recursive coarsening, an initial routing, and recursive refinement process presented in [85]. The major milestone of this model is the thermal-driven via planning algorithm. Based on this global view and capabilities of a multilevel planning scheme, the via planing step effectively optimize the temperature distribution and wire-length using direct planning of the inter-tier vias instead of indirect planning through a routing path search. This approach allows to control the chip temperature effectively. It also should be noted that any inter-tier via is also considered as a

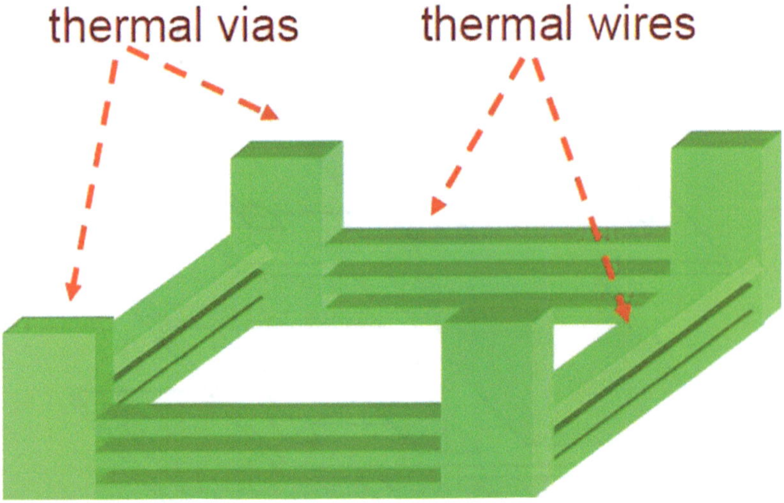

Fig. 2.15 An illustration of 3D thermal net designed to transfer heat vertically and horizontally from one location to the other

thermal via. The approach provides flexibility to add dummy inter-tier vias when the signal inter-tier vias number is not sufficient to bring down the chip temperature to an acceptable level. An initial routing solution is built using a 3D minimum spanning tree (MST) for each multi-pin net. Obstacles, such as thermal via regions, are avoided by using a simple maze routing algorithm. Then, the number of dummy inter-tier vias to be inserted is estimated using binary search. The upper bound of dummy inter-tier via numbers that can be inserted into each device layer is estimated by the amount of whitespace between the blocks. During each refinement stage, the inter-tier vias are refined first to minimize wire-length and temperature. This via-refinement process includes two steps, inter-tier via number distribution and signal inter-tier via assignment. These steps try to optimize temperature and wire-length, respectively. After inter-tier via refinement, wires are also adjusted according to the updated via positions Another approach is presented in [50]. It tackles the temperature-aware 3D routing problem not only by using thermal vias but also by introducing the concept of thermal wires. Thermal wires are objects with the function of spreading thermal energy in the lateral direction. Thermal vias perform the bulk of the conduction to the heat sink, while thermal wires help distributing the heat paths among multiple thermal vias. This strategy not only limits the temperature impact on the delay of signal nets, but also reduces congestion in hot regions. This is beneficial since more thermal vias may be inserted later into these regions to reduce the chip temperature. Figure 2.15 illustrate a model thermal net that is used in modern 3D chip design to transfer heat vertically and horizontally. The thermal nets are constructed using TTSVs and horizontal metal wires specially designed for heat transfer. The width and length of horizontal thermal wires can be bigger than horizontal signal wires.

2.6 3D-IC Design Verification

Many of the capabilities required for successful TSV design verification exist today in some EDA tools that are commercially available. There are, however a number of significant omissions and obstacles to avoid in *standard* physical verification methodologies. Some re-factoring of the tired and true design methodologies that are in place today is needed to accommodate TSV design structures as part of the verification landscape. The progress that take place EDA industry are evolutionary, rather than a revolutionary, approach in developing the 3D IC design tools. This appears to be a good decision because the technology, the rules and the standards are still evolving. The main EDA challenges are expected in the design space exploration [87], automatic across-die design partitioning, placement and routing, thermal and stress management, and 3D stack testing.

In the design space today, the location of each TSV is carefully orchestrated and tracked. A precise connection is made to each specific micro-bump on another die designed for the match, or other commodity chip with pre-defined locations. A commodity chip might be a memory, for example that is available from multiple sources, with same TSV pin-out and compatible characteristics. In such scenarios, TSV locations are deliberate and precise, acting more like an embedded connector than a traditional via. Regardless of the design complexities and the perceived need for new design methodologies for TSV design, there is still a fundamental need for a separate physical verification flow. An increase in complexity is the design domain does not necessarily have to follow into verification complexity increase of the same magnitude. Regardless of design style and methodology, physical verification is a necessary step to accurately verify design rule compliance, 3D stack LVS checking across the die parasitic extraction and simulation. Existing verification flows include the use of DRC, LVS and extraction to verify the connectivity of multiple stacked dies [88]. 3D-IC designs that utilize TSVs are essentially a double sided die. The TSV connects the regular front metal stack and back metal stack as illustrated in Fig. 2.16. The back metal stack provides for routing flexibility and consists of one or two layers. TSVs typically connect the first metals in the front (M1) and back metal (M6 or higher) stacks are manufactured using *via-first* process sequence and the TSVs go through the entire metal stacks are manufactured using *via-last* process sequence and result in smaller TSV densities since they require significant area to pass through the metal stacks. Various TSV models have been proposed ranging from piratically ignoring the TSVs altogether in verification (i.e. modeling them as single small resistance), to the complex models based of fitting S-parameter measurements with the parameter of TSV model [32]. Electrical characteristics of the TSV depend on its physical dimensions, size and thickness of its liner as well as on the material characteristics of silicon substrate. At the present stage of technology development and 3D-IC applications, it is assumed that there are no interactions between the TSVs, and a model for a single TSV is provided by the foundries. For verification purposes, TSVs are treated as an LVS device (GDS based flow) or as a Via (LEF/DEF based flow), and the provided spice TSV model of arbitrary complexity to be used for

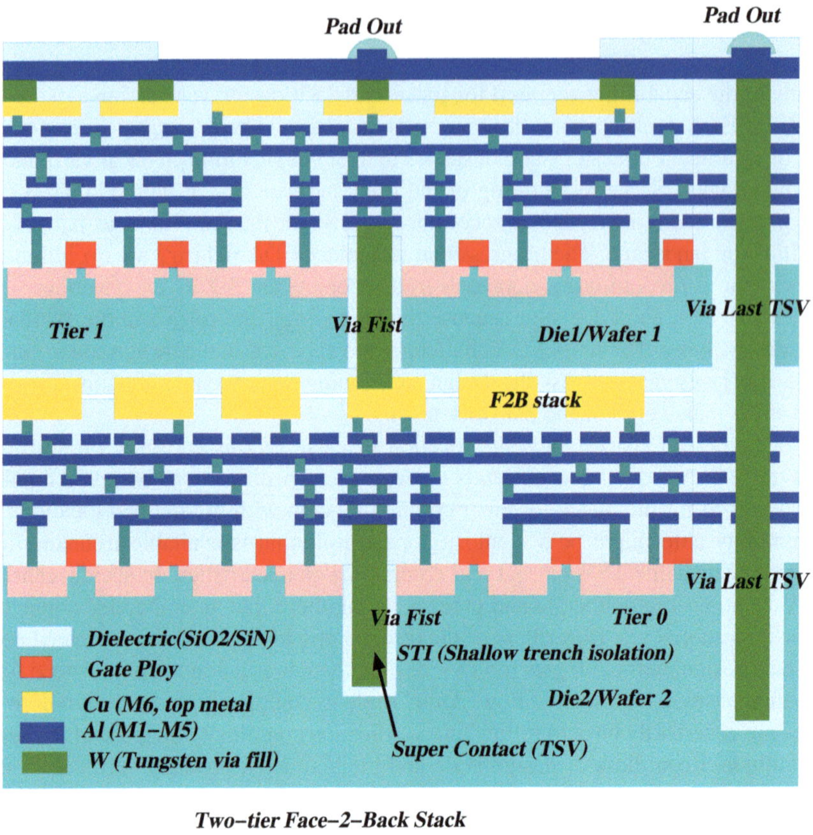

Fig. 2.16 Via first and via last with front and back metal stack

downstream simulation. This assumption, however, may not hold true as the technology advances and the TSV densities and frequencies become high, requiring accurate modeling and extraction of the interactions. Significant research effort has been put into this area recently [89–92] and it is expected to intensify in the coming years. This new methodology would require more accurate process description, accurate frequency dependent modeling and appropriate flow development to take advantage of the modeling accuracy.

2.7 Summary

Developing a design flow for 3D-ICs is a complicated task with many ramifications. Design methodologies at the frontend and mature manufacturing processes at the backend are required to effectively provide large scale 3D systems. Physical design techniques at different stages of a developmental design flow for 3D circuits have been discussed in this chapter, emphasizing the effect of the 3D nature on each design stage. A variety of floorplanning, placement, and routing techniques and algorithms for 3D circuits have been described that consider the unique characteristics of 3D circuits. In these techniques, the discrete nature of the third dimension is exploited to decrease the number of candidate solutions and, consequently, the computational time required to design a 3D circuit. Due to increased power densities and greater distances between the circuits on the upper planes and the heat sink, physical design techniques that embody a thermal objective can be a useful mechanism to manage thermal issues in 3D-ICs. Design techniques can reduce thermal gradients and temperatures in 3D circuits by redistributing the blocks among and within the planes of a 3D circuit. Alternatively, thermal vias can be utilized in 3D circuits to transfer heat to the heat sink. Thermal wires in the horizontal direction are similar in function to thermal vias and can also be utilized to lower thermal gradients within 3D circuits.

Another requirement for maximizing the speed of 3D circuits is to reliably distribute the clock signal within these circuits. A 3D clock distribution network, however, cannot be directly extended from a 2D circuit due to the asymmetry of a multi-tier 3D circuit and the effect of the inter-tier via impedance. Several clock distribution networks have been developed to investigate synchronization issues in 3D systems. In addition to higher performance, 3D integration offers significant opportunities for designing highly diverse and complex systems. Research on the design of 3D-ICs has only recently begun to emerge. Many challenges remain unsolved and significant effort is required to provide effective solutions to the problems encountered in the design of 3D-ICs. Furthermore, distributing power to the tiers of the stack located far from the power/ground pads is another fundamental issue in 3D-ICs. As the power/ground pads are typically located along the edges of the plane, providing sufficient current while satisfying target voltage levels for every transistor within a 3D-IC requires innovative power distribution networks. Addressing these important design issues will considerably accelerate the development of commercial 3D integrated systems.

References

1. G. Moore, Cramming more components onto integrated circuits. Proc. IEEE **86**(2), 82–85 (1998)
2. A. Rahman, A. Fan, J. Chung, R. Reif, Wire-length distribution of three-dimensional integrated circuits, in *Proceedings of the IEEE International Interconnect Technology Conference*, pp. 233–235, May 1999
3. A. Rahman, R. Reif, System level performance evaluation of three-dimensional integrated circuits. IEEE Trans. Very Large Scale (VLSI) Syst. **8**, 671–678 (2000)
4. D. Stroobandt, J. Van Campenhout, Accurate interconnection lengths in three-dimensional computer systems. IEICE Trans. Inform. Syst. Spec. Issue Phys. Des. Deep Sub-micron **10**(1), 99–105 (2000)
5. J.W. Joyner, Impact of three-dimensional architectures on interconnects in gigascale integration. IEEE Trans. Very Large Scale (VLSI) Syst. **9**, 922–928 (2001)

6. J.W. Joyner, P. Zarkesh-Ha, J.D. Meindl, A Stochastic global net-length distribution for a three-dimensional system-on-a-chip (3D-SoC), in *Proceedings IEEE International ASIC/SOC Conference*, pp. 147–151, Sep 2001
7. R. Zhang, K. Roy, C.-K. Koh, D.B. Janes, Stochastic interconnect modeling, power trends, and performance characterization of 3-D circuits. IEEE Trans. Elect. Devices **48**, 638–652 (2001)
8. J.W. Joyner, J.D. Meindl, Opportunities for reduced power distribution using three-dimensional integration, in *Proceedings of the IEEE International Interconnect Technology Conference*, pp. 148–150, June 2002
9. B.S. Cherkauer, E.G. Friedman, A unified design methodology for CMOS tapered buffers. IEEE Trans. Very Large Scale (VLSI) Syst. **3**, 99–111 (1995)
10. K. Banerjee, S.K. Souri, P. Kapour, K.C. Saraswat, 3D-ICs: A novel chip design paradigm for improving deep-submicrometer interconnect performance and systems-on-chip integration. Proc. IEEE **89**, 602–633 (2001)
11. M. Koyanagi et al., Future system-on-silicon LSI chips. IEEE Micro **18**, 17–22 (1998)
12. V.K. Jain, S. Bhanja, G.H. Chapman, L. Doddannagari, A highly reconfigurable computing array: DSP plane of a 3D heterogeneous SoC, in *Proceedings of the IEEE International System on Chip Conference*, pp. 243–246, Sep 2005
13. V.F. Pavlidis, E.G. Friedman, *Three-Dimensional Integrated Circuit Design Morgen Kaufmann* (2009). ISBN: 978-0-12-374343-5
14. R.J. Gutmann et al., Three-dimensional (3D) ICs: a technology platform for integrated systems and opportunities for new polymeric adhesives, in *Proceedings of IEEE International Conference on Polymers Adhesives Microelectron. Photon*, pp. 173–180, Oct 2001
15. M. Healy et al., Multiobjective microarchitectural floorplanning for 2-D and 3-D ICs. IEEE Trans. Comput. Aided Des. Integr. Circuits Syst. **26**, 38–52 (2007)
16. P. Garrou, C. Bower, P. Ramm, *Handbook of 3D Integration* (Wiley-VCH, New York, 2008). ISBN: 978-3-527-32034-9
17. R. Tummula, M. Swaminathan, *System-On-Package: Miniaturization of the Entire System* (McGraw-Hill, New York, 2008)
18. J.H. Lau, Critical issues of 3D IC integration, IMAPS transactions. J. Microelectron. Electron. Packag. (First Quarter Issue), 35–43 (2010)
19. J.H. Lau, Heart and Soul of 3D IC Integration, posted at 3D InCites on June 29 (2010), http://www.semineedle.com/posting/34277. Accessed 29 June 2010
20. K.W. Guarini, A.T. Topol, M. Ieong, R. Yu, L. Shi, M.R. Newport, D.J. Frank, Electrical integrity of state-of-the-art 0.13μm SOI CMOS devices and circuits transferred for Three-dimensional (3D) Integrated Circuit (IC) fabrication, in *IEDM Technical Digest*, pp. 943–945, 2002
21. R. Berridge, R.M. Averill III, A.E. Barish, M.A. Bowen, P.J. Camporese, J. DiLullo, P.E. Dudley, IBM POWER6 microprocessor physical design and design methodology. IBM J. Res. Dev. **51**(6), 685–714 (2007)
22. Y. Orii, T. Nishio, Ultra-thin POP technologies using 50 μm pitch flip chip C4 interconnections, in *Presented at the Electronic Components and Technology Conference (ECTC)* (Reno, NV, 2007)
23. J.J.Q. Lu, R. Gutmann, T. Matthias, P. Lindner, Aligned wafer bonding for 3-D interconnect, http://www.reed-electronics.com/semiconductor/article/CA630263. Accessed Aug 2005
24. K. Takahaski, Y. Taguchi, M. Tomisaka, H. Yonemara, M. Hoshino, M. Ueno, Y. Egawa, Process integration of 3D chip stack with vertical interconnection, in *Proceedings of the 54th Electronic Components and Technology Conference*, pp. 601–609, 1–4 June 2004
25. M. Umemoto, K. Tanida, Y. Nemoto, M. Hoshino, K. Kojima, Y. Shirai, K. Takahashi, High performance vertical interconnection for high-density 3D chip stacking package. in *Proceedings of the 54th Electronic Components and Technology Conference*, pp. 616–623, 1–4 June 2004
26. M. Feil, C. Adler, D. Hemmetzberger, M. Konig, K. Bock, The challenge of ultra thin chip assembly, in *Proceedings of the 54th Electronic Components and Technology Conference*, pp. 35–40, 1–4 June 2004

27. M. Hutter, F. Hohnke, H. Oppermann, M. Klein, and G. Engelmann, Assembly and reliability of flip chip solder joints using miniaturized Au/Sn bumps, in *Proceedings of the 54th Electronic Components and Technology Conference*, pp. 49–57 (2004)
28. V. Kripeshm, S. Yoon, S.W. Yoon, V.P. Ganesh, N. Khan, M.D. Rotaru, W. Fang, M.K. Iyer, Three-dimensional system-in-package using stacked silicon platform technology. IEEE Trans. Adv. Packag. **28**(3), 377–386 (2005)
29. H. Ikeda, M. Kawano, T. Mitsuhashi, Stacked memory chip technology development, in *SEMI Technology Symposium (STS) 2005 Proceedings, Session 9*, pp. 37–42 (2005)
30. S. Gupta, M. Hilbert, S. Hong, R. Patti, *Techniques for Producing 3D ICs with High-Density Interconnect* (Tezzaron Semiconductor Naperville, IL, 2005)
31. R. Patti, Advances in 3D memory and logic devices, in *IMAPS International Conference on Device Packaging, TAI3* (Scottsdale, AZ, 2010)
32. D. Min Jang, C. Ryu, K. Yong Lee, B. Hoon Cho, J. Kim, T. Sung Oh, W. Jong Lee, J. Yu, Development and evaluation of 3-D SiP with vertically interconnected Through Silicon Vias (TSV), in *Proceedings 57th Electronic Components and Technology Conference, ECTC-07*, Reno, NV, pp. 847–852 (2007)
33. M. Sadaka, I. Radu, L. di Cioccio, 3D Integration: advantages, enabling technologies and applications, in *IEEE International Conference on IC Design and Technology (ICICDT)*, Grenoble, France, pp. 106–109 (2010)
34. S.J. Koester et al., Wafer level 3D integration technology. IBM J. Res. Technol. IBM Res. Dev. **52**(6), 585–597 (2008)
35. D.E. Goldberg et al., *Genetic Algorithms in Search, Optimization, and Machine Learning* (Addison-Wesley, Reading, 1989)
36. A. Harter et al., *Three-Dimensional Integrated Circuit Layout* (Cambridge University Press, Cambridge, 1991)
37. C. Ryu et al., High frequency electrical circuit model of chip-to-chip vertical via interconnection for 3-D chip stacking package, in *Proceedings of IEEE Topical Meeting Electrical Performance of Electronic Packaging*, pp. 151–154, Oct 2005
38. D.M. Jang et al., Development and evaluation of 3-D SiP with Vertically Interconnected Through Silicon Vias (TSV), in *Proceedings of the IEEE International Electronic Components Technology Conference*, pp. 847–850, June 2007
39. V.F. Pavlidis, E.G. Friedman, Interconnect delay minimization through interlayer via placement in 3-D ICs, in *Proceedings of the ACM Great Lakes Symposium on VLSI*, pp. 20–25, Apr 2005
40. S. Tayu, S. Ueno, On the complexity of three-dimensional channel routing, in *Proceedings of the IEEE International Symposium on Circuits Systems*, pp. 3399–3402, May 2007
41. C. Addo-Quaye, Thermal-aware mapping and placement for 3-D NoC designs, in *Proceedings of the IEEE International SOC Conference*, pp. 25–28, Sep 2005
42. D. Hyun Kim, S. Mukhopadhyay, S. Kyu Lim, Through-silicon-via aware interconnect prediction and optimization for 3D stacked ICs, in *ACM/IEEE International Workshop on System Level Interconnect Prediction* (2009)
43. J.U. Knickerbocker et al., Three-dimensional silicon integration. IBM J. Res. Dev. **52**(6) (2008)
44. T.-Y. Chiang, S.J. Souri, C.O. Chui, K.C. Saraswat, Thermal analysis of heterogeneous 3D-ICs with various integration scenarios, in *Proceedings of IEEE International Electron Devices Meeting*, pp. 681–684, Dec 2001
45. C.C. Liu, J. Zhang, A.K. Datta, S. Tiwari, Heating effects of clock drivers in bulk, SOI, and 3D CMOS. IEEE Trans. Elect. Device Lett. **23**(12), 716–728 (2002)
46. G. Digele, S. Lindenkreuz, E. Kasper, Fully coupled dynamic electro-thermal simulation, IEEE Trans. Very Large Scale (VLSI) Syst. **5**, 250–257 (1997)
47. Z. Tan, M. Furmanczyk, M. Turowski, A. Przekwas, CFD-micromesh: a fast geometrical modeling and mesh generation tool for 3D microsystem simulations, in *Proceedings of the International Conference on Modeling Simulation Microsystems*, pp. 712–715, March 2000
48. P. Wilkerson, M. Furmanczyk, M. Turowski, Compact thermal model analysis for 3-D integrated circuits, in *Proceedings of the International Conference on Mixed Design Integration of Circuits Systems*, pp. 277–282, June 2004

49. M.B. Kleiner, S.A. Kahn, P. Ramn, W. Weber, Thermal analysis of vertically integrated circuits, in *Proceedings of IEEE International Electron Devices Meeting*, pp. 487–490, Dec 1995
50. T. Zhang, Y. Zhang, S. Sapatnekar, Temperature-aware routing in 3D-ICs, in *Proceedings of the IEEE Asia South Pacific Design Automation Conference*, pp. 309–314, Jan 2006
51. X. Zhao, D. Lewis, H.H.S. Lee, S. Kyu Lim, Pre-bond testable low-power clock tree design for 3D stacked ICs, in *IEEE International Conference on Computer-Aided Design* (2009)
52. J. Yang, K. Athikulwongse, Y.J. Lee, S. Kyu Lim, D. Pan, TSV stress aware timing analysis with applications to 3D-IC layout optimization, in *ACM Design Automation Conference* (2010)
53. R.H.J.M. Otten et al., Automatic floorplan design, in *Proceedings of IEEE/ACM Design Automation Conference*, pp. 261–267, June 1982
54. X. Hong et al., Corner block list: an effective and efficient topological representation of non-slicing floorplan, in *Proceedings IEEE/ACM International Conference on Computer-Aided Design*, pp. 8–11, Nov 2000
55. E.F.Y. Yong, C.C.N. Chu, C.S. Zion, Twin binary sequences: a non-redundant representation for general non-slicing floorplan. IEEE Trans. Comput. Aided Des. Integr. Circuits Syst. **22**, 457–469 (2003)
56. H. Yamazaki, K. Sakanushi, S. Nakatake, Y. Kajitani, The 3D-packing by meta data structure and packing heuristics. IEICE Trans. Fundam. Electron. Commun. Comput. Sci. **E83-A**(4), 639–645 (2000)
57. Y. Deng, W.P. Maly, Interconnect characteristics of 2.5-D system integration scheme, in *Proceedings of the IEEE International Symposium Physical Design*, pp. 341–345, Apr 2001
58. L. Cheng, L. Deng, D.F. Wong, Floorplanning for 3-D VLSI design, in *Proceedings of the IEEE International Asia South Pacific Design Automation Conference*, pp. 405–411, Jan 2005
59. Z. Li et al., Hierarchical 3D floorplanning algorithm for wire length optimization. IEEE Trans. Circuits Syst. I Regul. Pap. **53**(12), 2637–2646 (2006)
60. P. Zhou, Y. Ma, Z. Li, R. Dick, L. Shang, H. Zhou, X. Hong, Q. Zhou, 3D-STAF: scalable temperature and leakage aware floorplanning for three-dimensional integrated circuits, in *Proceedings of the ICCAD*, pp. 590–597 (2007)
61. J. Cong, J. Wie, Y. Zhang, A thermal-driven floorplanning algorithm for 3D-ICs, in *Proceedings of ICCAD*, pp. 306–313 (2004)
62. L. Cheng, L. Deng, M.D.F. Wong, Floorplanning for 3D-VLSI design, in *IEEE International Asia South Pacific Design Automation Conference (ASPDAC)*, pp. 405–411 (2005)
63. M.W. Newman et al., Fabrication and electrical characterization of 3D vertical interconnects, in *Proceedings of the IEEE International Electronic Components Technology Conference*, pp. 394–398, June 2006
64. W.-C. Lo et al., An innovative chip-to-wafer and wafer-to-wafer stacking, in *Proceedings of the IEEE International Electronic Components Technology Conference*, pp. 409–414, June 2006
65. B. Goplen, S. Sapatnekar, Placement of thermal vias in 3D-ICs using various thermal objectives. IEEE Trans. Comput. Aided Des. Integr. Circuits Syst. **25**, 692–709 (2006)
66. M. Ohmura et al., An initial placement algorithm for 3D VLSI, in *Proceedings of IEEE International Symposium on Circuits Systems*, vol. IV, pp. 195–198, May 1998
67. T. Tanprasert et al., An analytical 3D placement that preserves routing space, in *Proceedings of the IEEE International Symposium on Circuits Systems*, vol. III, pp. 69–72, May 2000
68. Y. Deng, W.P. Maly, Interconnect characteristics of 2.5D system integration scheme, in *Proceedings of the ACM International Symposium on Physical Design*, pp. 171–175, Apr 2001
69. I. Kaya, M. Olbrich, E. Barke, 3D Placement considering vertical interconnects, in *Proceedings of the IEEE International SOC Conference*, pp. 257–258, Sep 2003
70. S.T. Obenaus, T.H. Szymanski, Gravity: fast placement for 3-D VLSI. ACM Trans. Des. Autom. Electron. Syst. **8**(3), 298–315 (2003)
71. W.R. Davis et al., Demystifying 3D-ICs: the pros and cons of going vertical. IEEE Des. Test Comput. **22** (2005)
72. H. Eisenmann, F.M. Johannnes, Generic global placement and floorplanning, in *Proceedings of IEEE/ACM Design Automation Conference*, pp. 269–274, June 1998

73. B. Goplen, S. Sapatnekar, Efficient thermal placement of standard cells in 3D-ICs using a force directed approach, in *Proceedings of the IEEE/ACM International Conference on Computer-Aided Design*, pp. 86–89, Nov 2003
74. MCNC Benchmarks, http://er.cs.ucla.edu/benchmarks/ibm-place
75. IBM-PLACE Benchmarks, http://www.cbl.ncsu.edu/pub/Benchmark_dirs/LayoutSynth92
76. B. Black et al., Die stacking (3D) microarchitecture, *Proceedings of IEEE/ACM International Symponsium on Micro-architecture*, pp. 469–479, Dec 2006
77. B. Goplen, S. Sapatnekar, Thermal via placement in 3D-ICs, in *ISPD*, pp. 167–174 (2005)
78. Z. Li et al., Efficient thermal via planning approach and its application in 3D floorplanning. IEEE Trans. Comput. Aided Des. Integr. Circuits Syst. **26**, 645–658 (2007)
79. R.J. Enbody, G. Lynn, K.H. Tan, Routing the 3D chip, in *Proceedings of IEEE/ACM Design Automation Conference*, pp. 132–137, June 1991
80. C.C. Tong, C. Wu, Routing in a three-dimensional chip. IEEE Trans. Comput. **44**(1), 106–117 (1995)
81. J. Minz, S.K. Lim, Block-level 3D global routing with an application to 3D packaging. IEEE Trans. Comput. Aided Des. Integr. Circuits Syst. **25**, 2248–2257 (2006)
82. A. Hashimoto, J. Stevens, Wire routing by optimizing channel assignment within large apertures, in *Proceedings of IEEE/ACM Design Automation Conference*, pp. 155–169, June 1971
83. T. Ohtsuki, E. HorbstT, *Advances in CAD for VLSI: Logic Design and Simulation* (The University of Michigan, North-Holland, 1986). ISBN: 444878920, 9780444878922
84. J. Cong, M. Xie, Y. Zhang, An enhanced multilevel routing system, in *Proceedings of the IEEE/ACM International Conference on Computer-Aided Design*, pp. 51–58, Nov 2002
85. J. Cong, Y. Zhang, Thermal driven multilevel routing for 3D-ICs, in *Proceedings of the IEEE Asia and South Pacific Design Automation Conference*, pp. 121–126, June 2005
86. J. Cong, Y. Zhang, Thermal via planning for 3D-ICs, in *Proceedings of the IEEE/ACM International Conference on Computer-Aided Design*, pp. 744–751, Nov 2005
87. D. Milojevic, R. Radojcic, R. Carpenter, P. Marchal, Pathfinding: a design methodology for fast exploration and optimisation of 3D-stacked integrated circuits, in *International Symposium on System-on-Chip, 2009. SOC 2009*, pp. 118–123 (2009)
88. M. Hogan, D. Petranovic, Robust verification of 3D-ICs: Pros, Cons and Recommendations, in *IEEE International Conference on 3D System Integration, 2009. 3DIC*, pp. 1–6, 28–30 Sept 2009
89. J.H. Wu, Through-substrate Interconnects for 3-D Integration and RF systems. Ph.D. dissertation, MIT, Cambridge, MA, Oct 2006
90. I. Savidis, E.G. Friedman, Electrical modeling and characterization of 3-D vias, in *Proceedings of the IEEE International Symposium on Circuits Systems*, pp. 784–787, May 2008
91. I. Savidis, E.G. Friedman, Closed-form expressions of 3-D via resistance, inductance,and capacitance. IEEE Trans. Electron Devices **56**(9), 1873–1881 (2009)
92. D. Hyun Kim, K. Athikulwongse, S. Kyu Lim, A study of through-silicon-via impact on the 3D stacked IC layout, in *IEEE International Conference on Computer-Aided Design* (2009)

Chapter 3
Field Programmable Gate Arrays: An Overview

Abstract Field Programmable Gate Arrays (FPGAs) are semiconductor devices that contain logic components connected by a regular, hierarchical programmable interconnect system. The distinguishing characteristic of FPGAs is their on-filed programmability which allows the logic functionality of an FPGA to be re-programmed even after the manufacturing process. FPGAs are used for rapid prototyping of digital circuits. The design and test of digital systems are time efficient and cost-effective with FPGAs. The logic components in the FPGA mostly consists of memory elements such as registers or even complete blocks of memory that can be configured to hold any desired state. The hierarchical interconnect system is also programmable which allows the logic components to be connected in a variety of network configurations. Therefore the re-programmability of FPGAs is achieved by a fixed underlying architecture, which does not cater to any particular logic circuit. This lets FPGAs have a lower non-recurring cost, shorter design cycle and enables them to be re-programmed in the field to circumvent manufacturing defects. This chapter discuses about the FPGA building blocks and how they are interconnected to form a flexible digital prototyping and design platform.

3.1 Introduction

Field Programmable Gate Arrays (FPGAs) [1–6] are semiconductor devices that contain logic components connected by a regular, hierarchical programmable interconnect system. The distinguishing characteristic of FPGAs is their on-filed programmability which allows the logic functionality of an FPGA to be re-programmed even after the manufacturing process. FPGAs are used for rapid prototyping of digital circuits. The design and test of digital systems are very time-efficient and cost-effective with FPGAs. The logic components in the FPGA mostly consists of memory elements such as registers or even complete blocks of memory that can be configured to hold any desired state. The hierarchical interconnect system is also programmable which allows the logic components to be connected in a variety of network configurations. Therefore the re-programmability of FPGAs is achieved by a fixed underlying architecture, which does not cater to any particular logic circuit. This lets FPGAs have a

V. Pangracious et al., *Three-Dimensional Design Methodologies for Tree-based FPGA Architecture*, Lecture Notes in Electrical Engineering 350, DOI 10.1007/978-3-319-19174-4_3

lower non-recurring cost, shorter design cycle and enables them to be re-programmed in the field to circumvent manufacturing defects. FPGS were used mostly for prototyping and emulation systems in the design process for ASICs. However, recently, FPGAs have become popular for a variety of mainstream products in networking, telecommunication, digital signal processing and in consumer electronics. FPGAs can be classified based on the technology using to program it.

- Antifuse FPGAs—The devices are configured by burning a set of fuses. Once the chip is configured, it cannot be reprogrammed any more.
- Flash FPGAs—These devices can be reprogrammed several thousands of times and retain their configuration even when the power is switched off. However, these devices take several seconds for reconfiguration.
- SRAM based FPGAs—These devices are most popular kind of devices as they offer unlimited re-programming using Static Random Access Memory (SRAM) cells to hold the circuit configuration that is loaded into the FPGA. They offer very fast reconfiguration, with some devices such as the Xilinx [7] allowing even partial reconfiguration.

3.2 Introduction to FPGA Architectures

FPGAs consists of an array of configurable logic blocks (LBs) that are connected by a 2D grid of metal channels, as illustrated in Fig. 3.1. Each LBs contains a small amount of digital logic to form a Look-Up-Table (LUT) which implement the boolean logic functions using SRAM bit cells. A n bit SRAM can be used to implement any one of the 2^{2n} boolean functions that have n inputs and a single output. The output of the LUTs are also connected to registers whose outs can be chosen instead of the direct LUT output. Thus a LB can be programmed by a small amount of memory to implement sequential logic as well as combinational logic. The mesh-based channels connecting these LBs together contains switch boxes (SBs) at the grid-points, which can connect the intersecting channels to each other using programmable switches. The programmable memory in the LBs as well as the memory controlling switches in the SBs, together form the configuration memory of FPGA. FPGA is a prefabricated silicon device that can be re-configured to implement various applications. Re-configurability is derived from re-programmable SRAMs. By programming the SRAM cells, the functionality of FPGA logic blocks can be tailored to implement a particular logic circuit. LBs and interconnects are established by programming SRAM cells to connect prefabricated routing resources. Thus, any given logic circuits can be mapped into the FPGA by programming functionality and connectivity of logic blocks based on the specific characteristics of the application. Therefore any logic circuit is functionally equivalent to a particular state of the configuration memory of the FPGA. For this particular state of the configuration memory, The FPGA behaves exactly like the mapped circuit for all possible inputs of the circuit. The major challenge in FPGA design is to provide the maximum placement and routing

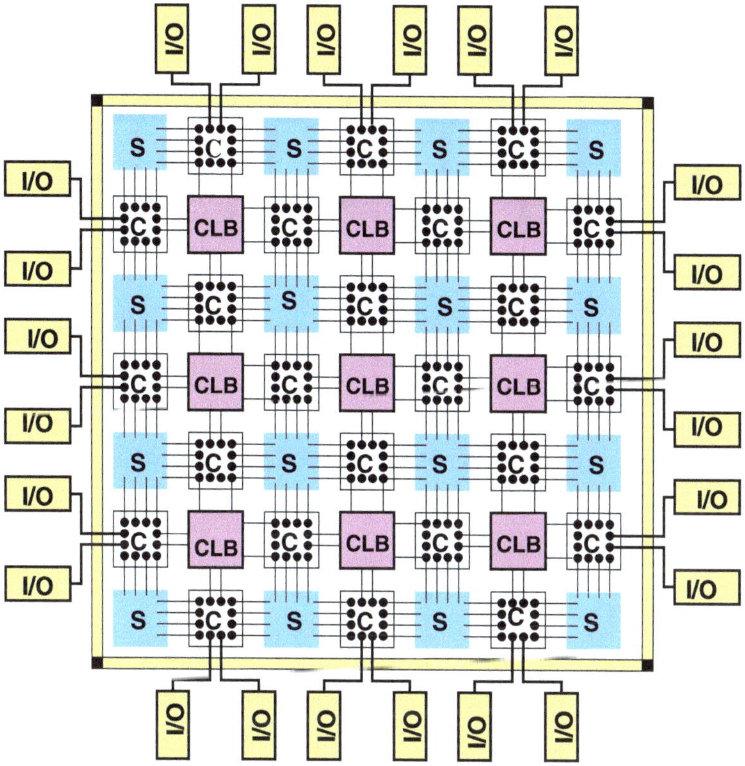

Fig. 3.1 A Mesh-based homogeneous FPGA architecture

flexibility with minimum area cost. FPGA designers propose different architecture and interconnection topologies to achieve this trade-off.

3.2.1 Configurable Logic Blocks

The Configurable Logic Block (CLB) as shown in Fig. 3.2 is a basic component of FPGA, contains the boolean logic that are the equivalent of gates in an ASIC circuit. A CLB comprise of a single basic logic element (BLE), or a cluster of locally interconnected BLEs. The BLE consists of Look-Up-Table (LUT) and Flip-Flop as illustrated in Fig. 3.3. A LUT with k inputs (LUT-k) contains 2^k configuration bits, capable of implementing any *k-input* boolean function. Figure 3.3 shows the BLE comprising of a four input Look-Up-Table (LUT-4) and D-type Flip-Flop. The LUT-4 uses 16-bit SRAM accesses by a 4-to-1 MUX tree to implement any four-inputs boolean function. The output of BLE is also connected to a Flip-Flop and a 2-to-1 multiplexer is used to choose between the direct BLE output and the latched output.

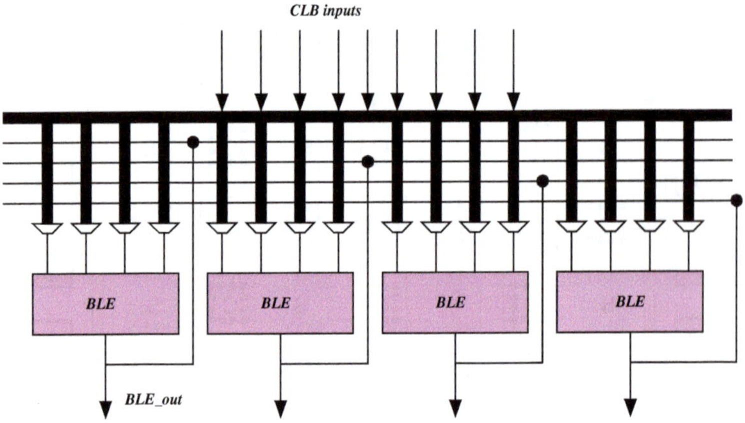

Fig. 3.2 Illustration of typical Configurable Logic Block of FPGA architecture

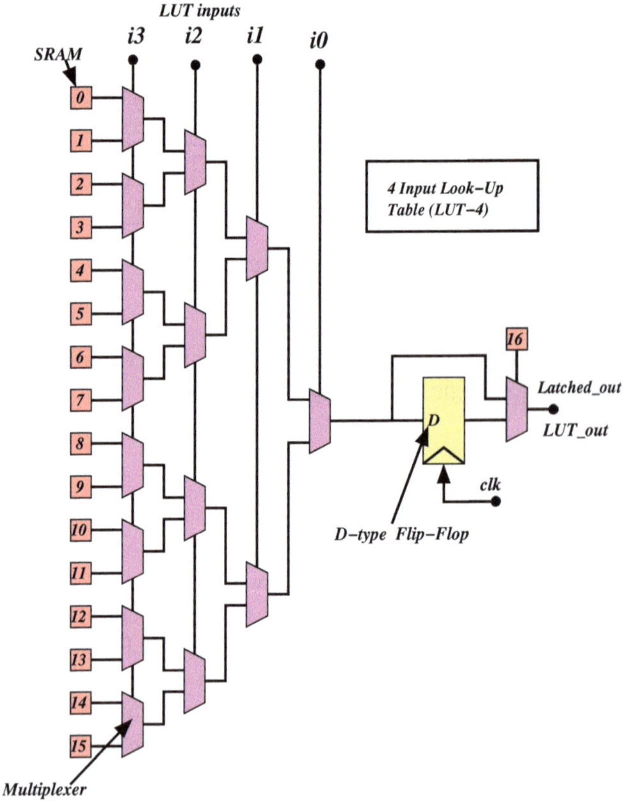

Fig. 3.3 An illustration of typical FPGA Look-up-Table (LUT) architecture

The latched output is used to implement sequential circuits in the FPGA. A BLE with more number of number of inputs reduces the total number of BLEs required to map a hardware circuit. This eventually reduces the intercommunication between BLEs and thus the speed of the hardware circuit improves. However the area increases exponentially as the number of inputs increases [8].

A CLB can contain a cluster of BLEs connected through the logically equivalent local routing network. Figure 3.2 shows a cluster of four BLEs and each BLE contains a LUT-4 and a Flip-Flop. The BLE also has internal routing between the logic block outputs and the inputs, which allows the BLE outputs to be mixed along with inputs. This feature is later exploited in the packing stage, where a chain of LUTs in a boolean network can be packed together into a logic block. This internal routing circuitry allows such BLEs to aviod routing through switch boxes, reducing the delay between BLEs. The number of outputs pins of a cluster are equal to the total number of BLEs in a cluster, whereas the number of inputs pins of a cluster can be less than or equal to the sum of input pins required by all the BLEs in the cluster. Modern FPGAs contain typically 4 to 10 BLEs in a single cluster. A cluster of BLEs is an alternative way to improve the granularity of an FPGA logic block. This approach is now commonly used in most industrial FPGAs. Here, several BLEs are grouped into cluster size N. By increasing the CLB granularity in this manner as opposed to growing it by making the BLE size larger, the size of the logic and internal routing to supply the complete crossbar connectivity within the cluster only grows quadratically versus exponential growth in area as BLE size increases [8]. In general, there are fewer inputs to the cluster from the external inter-cluster routing than the total number of inputs to the BLEs inside the cluster. This reduction is possible because the cluster input signals are often used as inputs to multiple BLEs in cluster with sufficient number of BLEs. This observation is first captured in an equation for clusters of N 4-input LUTs by [5] I, the number of pins needed to fully occupy a cluster of N 4-input LUTs is $2N + 2$, as opposed to the total number of inputs pins on all basic logic elements, $4N$. This relationship is generalized in [8] for *N-sized* clusters of k-inputs LUTs is shown in Eq. 3.1, which is significantly less than the maximum of KN

$$I = \frac{K}{2}(N + 1) \tag{3.1}$$

Typically, intra-cluster routing in contemporary BLE-based FPGAs does not exhibit full crossbar connectivity. The FPGA optimization study presented in [9] concluded that at least half of the connections between cluster inputs and logic elements inputs can be removed and between 50 and 75 % of the feedback connections from logic elements inputs can be removed with no impact on critical path delay or the number of logic clusters required. This switch and interconnect depopulation results in about a 10 % area reduction of FPGAs with cluster size similar to commercial FPGAs. The reduction in intra-cluster routing flexibility requires an FPGA router to extend the search for wiring paths into logic clusters. This extended search can increase routing time by up to factor of four [9].

3.3 FPGA Interconnect Topologies

In the design and development of FPGAs, we have studied Mesh-based [1–6], Tree-based [10–14] and unified Mesh of Tree interconnection network [15–17]. We have seen numerous research and studies showing the characteristics of these networks, how they scale, and empirically how they relate on particular designs [10, 12, 18]. We decided to revisit all tree-based architectures and interconnect network to study and understand their capabilities, which will help us to determine the changes and modifications required to make them suitable for sub-nanometer designs.

3.3.1 Mesh-Based Interconnect Network

The CLBs of an island styled Mesh-based FPGA are arranged in a two dimensional mesh with routing resources evenly distributed throughout the mesh as illustrated in Fig. 3.4. A planar global routing architecture typically has routing channels on all four sides of the CLBs. The number of wires contained in the channel, *W*, is pre-set during fabrication, and is one of the key choices made by the FPGA architect. The Mesh-based routing resources generally use wire segments of different lengths in each channel in an attempt to provide the most appropriate length for each given connection. Currently most commercial SRAM-based FPGA architectures [7] use Mesh-based FPGA architectures. The routing architecture of Mesh-based FPGA

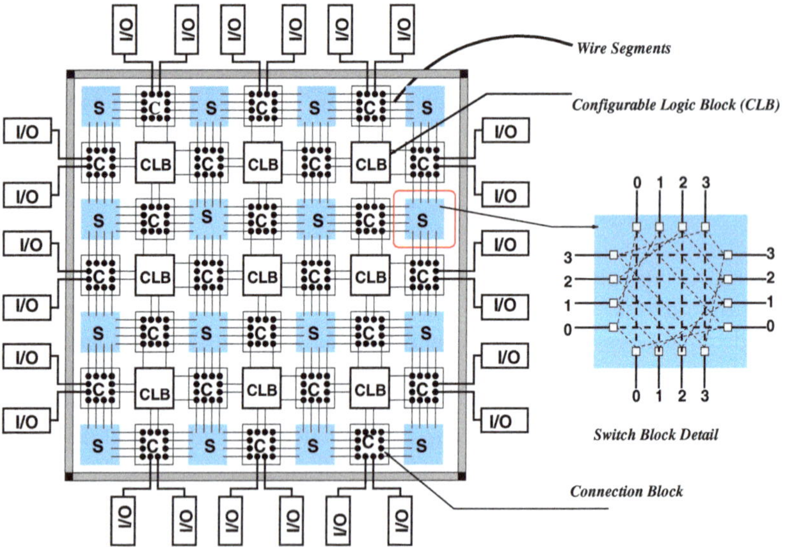

Fig. 3.4 An illustration of typical Mesh-based FPGA architecture and switch block (SB)

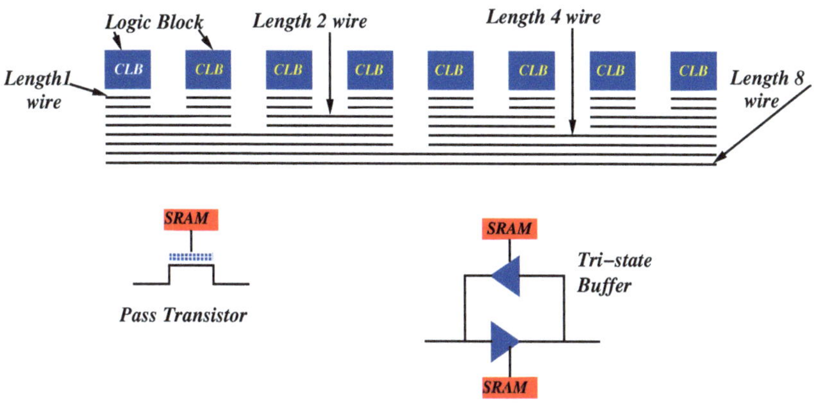

Fig. 3.5 Two types of programmable switches used in SRAM-based FPGAs

defines the logical structure of programmable interconnection between wire segments in routing channels and between logic block I/O and routing channel wire segments organized in rows and columns. The set of switches used to connect a logic block to an adjacent routing channel is called a connection block C. Similarly, the set of switches used to connect intersecting routing channels is called a switch block S. Each routing channel contains W parallel wires tracks, where W is called the channel width. The same width is used for all channels. Figure 3.5 illustrates these various routing structures. The structure of these individual routing components can be parameterized by routing channel width, segment distribution, connection block topology, and switch block topology. Segment distribution describes the lengths of the wire segments in the routing channels. Longer wire segments span multiple blocks and require fewer switches, thereby reducing routing area and delay. However, they also decrease routing flexibility, which reduces the probability that a user circuit can be routed successfully.

3.3.2 FPGA Switch Block

Modern FPGAs commonly use a combination of long and short wires in order to balance the routing area and delay tradeoff. Connection and switch block topologies describe the interconnection pattern within these blocks. In terms of routability, fully populated blocks as illustrated in Fig. 3.4 (that is, blocks for which any incident pin can be connected to any other incident pin) would be optimal. However, in terms of area, the cost would be prohibitive. Previous work [9, 19] has shown that connection and switch blocks still provide good routability even when only sparsely populated. Connection block population is defined by $F_{c_{in}}$ and $F_{c_{out}}$ parameters, where $F_{c_{in}}$ is routing channel to cluster input switch density and $F_{c_{out}}$ is cluster output to the routing channel density. The regular grid of routing channels is connect at the

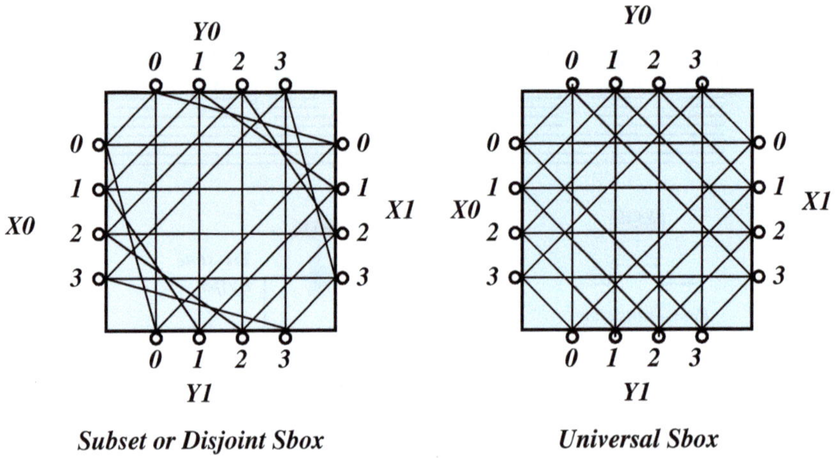

Fig. 3.6 Disjoint and universal 2D switch boxes, X_0, X_1, Y_0, Y_1 mark their sides

grid-points by a collection of switches know as a *switch matrix* or *switch block* (SB). A switch block is a set of programmable switches between the metal channels that can be configured to create any desired routing pattern. A switch block from the Mesh-based FPGA grid connect four metal channels, an $X0$ channel that connects from the left of the switch box, an $X1$ channel that connects from the right, a $Y0$ channel that connects from above and a $Y1$ channel that connect to the bottom of the switch block as illustrated in Fig. 3.6. Each of these channels usually has the same number of tracks W, which is know as *channel width* of the FPGA. Therefore each side of the switch block has W pins that have to be connected to each other is some routing pattern. The $switch\ flexibility$ F_s of a pin on some side of a switch block is the number its connections to pins on the remaining sides of the switch block. Figure 3.6 shows two different switch routing pattern with $F_s = 3$.

A popular switch block architecture is the $disjoint$ switch architecture which is shown in Fig. 3.6. In Fig. 3.6, it can be seen that when the pins of any side of the switch block are numbered from 1 to 4, since in this example the W is set to 4, in general a pin numbered i on one side of the switch block connect only to pins number i on the other three sides of the switch block. Hence the constraint on the routing pattern is that no pin i on one side can connect to a pin j on another side for $i \neq j$. Another example of a switch block architecture is the $universal$ switch block where every set of nets satisfying the dimensional constraint, which is the maximum number of W tracks in each channel, is simultaneously routable through the switch block. The choice of the switch block architecture can have a significant impact on the routability of the FPGA and consequently the performance of the FPGA. the connections between the pins of the switch blocks can be made through a pair of tristate buffers or through a bidirectional pass transistor as shown in Fig. 3.5. Programmable SRAM-based switches within connection blocks and switch blocks can be implemented using either pass-transistors or tri-state buffers, as illustrated in Fig. 3.5. Pass-transistor switches require less area and dissipate less power than tri-state buffer switches. However, tri-state buffer switches are faster for connections that span many segments. It is well known by VLSI designers that propagation delay through one pass transistor is smaller than corresponding delay through one buffer [20]. However, it is also known that placing many pass transistors in series is much slower than a similar chain of buffers because delay grows quadratically with the former, but linearly with the latter. Routing architectures commonly use a combination of tri-state buffer and pass-transistor switches to reduce area and delay. Global networks, such as clock and reset networks, are implemented with dedicated routing tracks which are separated from the configurable routing. Like other integrated circuits, FPGA clock distribution networks are designed to minimize skew in order to maximize system performance.

Any pin in the switch block can be driven by at most one out of the 3 switches that connect to it. The parasitic capacitance seen by this driving switch depends on the type of the switches that connect to this pin. When the connects are made through tristate buffers, the contribution made by each bidirectional connection to the capacitance seen by the driving pin is the sum of the input capacitance C_{in} and output capacitance C_{out} of the tristate buffer. Therefor capacitance encountered by a net driver at a switch block pin due the bidirectional switches alone, is given by Eq. 3.2.

$$C_{pin} = 3 \times (C_{in} + C_{out}) \tag{3.2}$$

The capacitance values is seen by the driving switch regardless of whether the other tristate buffer connections are open or closed. When pass transistor are used to make the connections, the state of the other connections, either open or closed, determines the capacitance seen be the driving switch. In other words, in the case of pass transistor connection, the fanout at a pin determines the parasitic capacitance seen by the switch driving the pin. The contribution of an open pass transistor switch is only the overlap capacitance value of the transistor. On the other hand, a closed transistor switch, apart from the closed transistor capacitance, exposes the entire capacitance tree behind this switch, up to the points where buffers are encountered.

3.3.3 FPGA Routing Channels

The metal channels that form the interconnect of the Mesh-based 2D FPGA consist of two type of channels, namely the X and Y channels. Each channels contains wire tracks that can be individually driven by a source. The wire in each track is broken up into a multiple wire segments that the same characteristics. It is found to be more beneficial to use segments of different lengths rather than a single uniform length through the FPGA [6]. This allows the router connecting two LBs in the FPGA, to choose the type of segment depending on the physical distance between the two LBs. *Long wire* segments that can *span* multiple switch blocks, bypassing intermediates the switch blocks, are ideal for connecting LBs that are far apart in the FPGA. On the other hand, to connect LBs that are very close to each other, long wire segments with higher wire capacitance, result in greater delay compared to shorter wire segments. A wire segment is characterized by the following properties.

1. Length—The wirelength of the segment is measured in unites of the distance between two adjacent switch blocks. For example a wire segment of span 2 connects two switch blocks that are separated by a single switch block. A segment of span 4, connects two switch blocks in a row or column, separated by three switch blocks.
2. LB population—When a wire segments span across M switch blocks and is connected to the corresponding LBs in only M out of the N switch blocks, the LB population is given by $\frac{M}{N}$.
3. Switch population—For a wire segment span across N switch blocks and connects through the switches to only M out of them, the switch population is given by $\frac{M}{N}$.

The different types of wire segments used in the Mesh-based 2D FPGA are summarized in Table 3.1. Thus a metal channel can be summarized by the following properties

1. Channel width W, which is the number of wire tracks within the channel.
2. Segment types—The different types of metal wire segments that form the channel.
3. Segment distribution—The proportion of each of these segments types in the channel and how these segments are interleaved.
4. Segment staggering—The optimal orientation of the start point of each of these segments types for maximum routability.

FPGA vendors do not offer FPGAs with different amounts of interconnects, for a given logic capacity. This is surprising since interconnect consumes nearly 90 % of the chip area. Some reasons for not offering a variety of interconnect sizes are inventory control, the impact of marketing and sales of inferior or unroutable devices, and the large amount of engineering effort required to develop a single device. The LUT size, the number of LBs in each cluster and the number of inputs per cluster vary with each vendor. For all experiments performed in this work, those parameters are chosen to be consistent with previous work [8]. Note that the channel width of the FPGA is left as a variable. The CAD tools developed for this work attempt to find the

Table 3.1 Mesh-based 2D FPGA segments types

Length	Number of spanned SBs	LB population	Number of connected LBs	SB population	Number of connected SBs
1	2	1.0	2	1.0	2
2	3	1.0	3	0.66	2
4	5	0.6	3	0.4	2

minimum possible channel width required to route a specific circuit. The interconnect requirement is tailored for the circuit to be implemented. This technique allows us to compare different interconnect topologies in terms of routability targeting different applications domains.

3.3.4 Multilevel Hierarchical Interconnect

Majority of the logic designs exhibit locality of connections implying a hierarchy in placement and routing of connections between logic blocks. The Hierarchical FPGA architecture attempts to exploit this feature to provide smaller routing delays and more predictable timing behavior. The speed of a net is determined by the number of routing switches it has to pass through. In a Mesh structure, the number of segments in series increases linearly with manhattan distance d, between the logic blocks to be connected. However in the case of Tree connectivity is that the number of switches in series in a route connecting two logic blocks increases as a logarithmic function of the manhattan distance. This is illustrated in Fig. 3.7. Multilevel hierarchical Interconnect regroups the interconnect architectures with more than 2 levels of hierarchy and Tree-based ones. For example VPR and APEX architectures are not included in this category since they have only 2 levels of hierarchy. Multilevel hierarchical architecture is created by connecting logic blocks into clusters. These clusters are recursively connected to form a hierarchical structure. In the hierarchical FPGA called HFPGA, LBs are grouped into clusters. Clusters are then grouped recursively together as

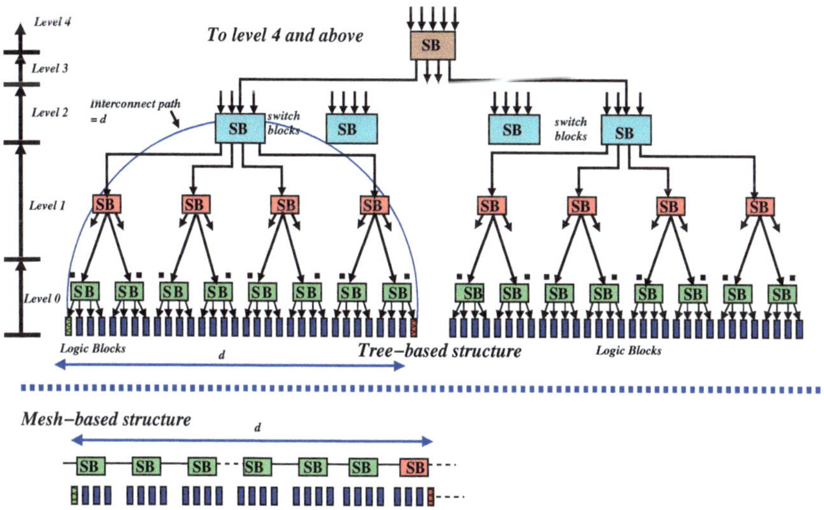

Fig. 3.7 Mesh vs Tree interconnect structure, number of series switched in Tree-based and Mesh-based interconnect

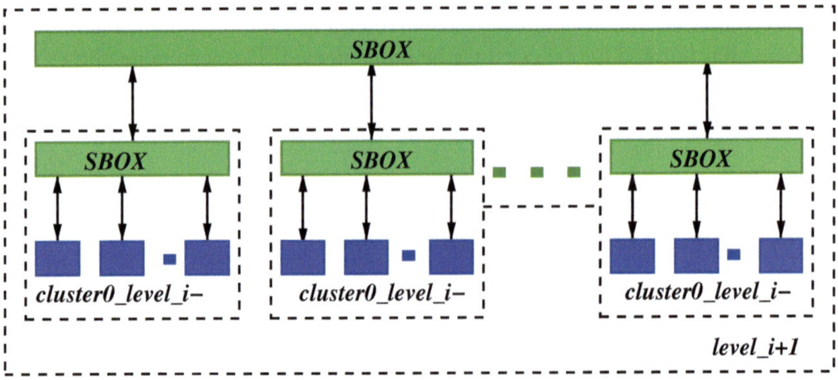

Fig. 3.8 Hierarchical FPGA topology

illustrated in Fig. 3.8. The clustered VPR mesh architecture has a Hierarchical topology with only two levels. Here we consider multilevel hierarchical architectures with more than 2 levels. In [10, 12] various hierarchical structures were discussed. The HFPGA routability depends on switch boxes topologies. HFPGAs comprising fully populated switch boxes ensure 100 % routability but are very penalizing in terms of area. In [12] authors explored the HFPGA architecture, investigating how the switch pattern can be partly depopulated while maintaining a good routability.

A well-known academic hierarchical FPGA is the Hierarchical Synchronous Reconfigurable Array (HSRA) [13]. HSRA has a strictly hierarchical Tree-based interconnect structure. Consequently, HSRA's logic and interconnect structures are not as closely coupled as the logic and interconnect structures of island-style FPGAs. In HSRA, the only wire-segments that directly connect to the logic units are located at the leaves of the interconnect tree. All other wire-segments are decoupled from the logic structure. A HSRA logic unit consists of a single 4-LUT/D-FF pair. The input-pin connectivity is based on a choose-k strategy [13], and the output pins are fully connected. The richness of HSRA interconnect structure is defined by its base channel width and interconnect growth rate. The base channel width c is the number of tracks at the leaves of the interconnect Tree as illustrated in Fig. 3.9 with $c = 3$. Growth rate p is the rate at which the interconnect grows towards the root (in Fig. 3.9, $p = 0.5$). The growth rate is realized using the following types of switch-blocks:

- Non-compressing (2:1) switch blocks—The number of root-going tracks is equal to the sum of the number of root-going tracks of the two children.
- Compressing (1:1) switch blocks—The number of root-going tracks is equal to the number of root-going tracks of either child.

A repeating combination of non-compressing and compressing switch blocks can be used to realize any value of p less than one. For example, a repeating pattern of (2:1 1:1) switch blocks realizes $p = 0.5$, while the pattern (2:1 2:1 1:1) realizes $p = 0.67$. A HSRA that has only 2:1 switch blocks provides maximum interconnection bandwidth (i.e. a value of $p = 1$). $APEX$ architecture is a commercial product from Altera Corporation which includes 3 levels of interconnect hierarchy. Figure 3.10

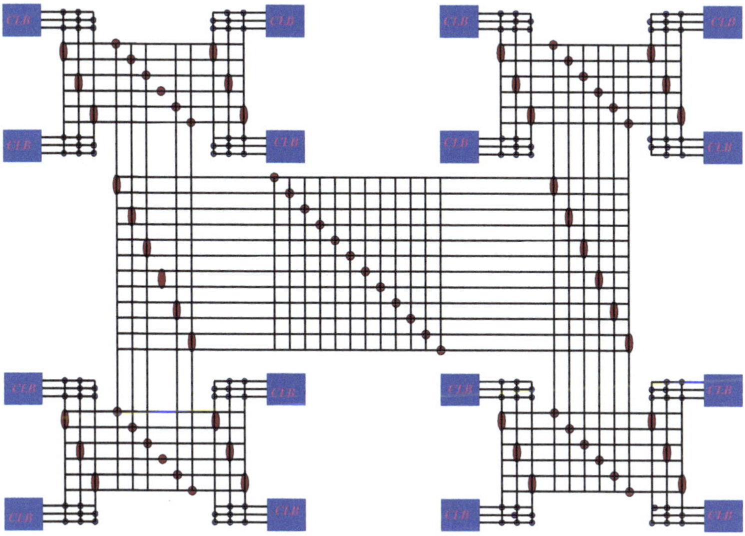

Fig. 3.9 HSRA interconnect structure

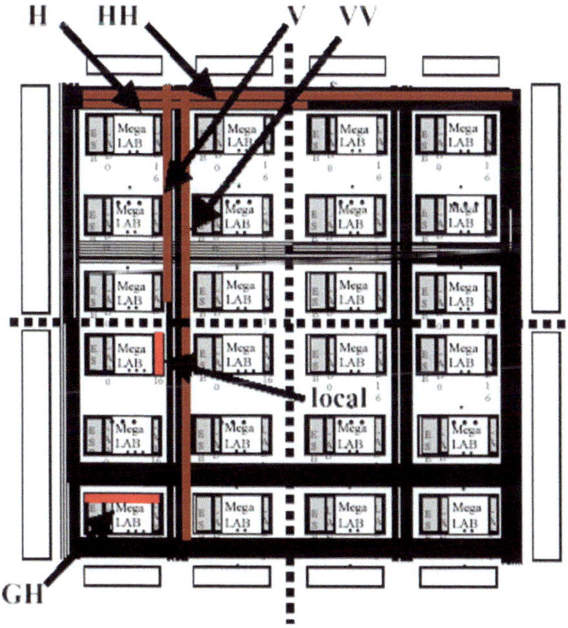

Fig. 3.10 The APEX programmable logic devices [21]

shows a diagram of the APEX 20K400 programmable logic device. The basic logic-element (LE) is a 4-input LUT and DFF pair. Groups of 10 LEs are grouped into a logic-array-block or LAB. Interconnect within a LAB is complete, meaning that a connection from the output of any LE to the input of another LE in its LAB always exists, and any signal entering the input region can reach every LAB. Groups of 16 LABs form a MegaLAB. Interconnect within a MegaLAB requires an LE to drive a global horizontal line or GH (MegaLAB global H) line, which switches into the input region of any other LAB in the same MegaLAB. Adjacent LABs have the ability to interleave their input regions, so any LE in LAB_i can usually drive LAB_{i+1} without using a GH line. A 20K400 MegaLAB contains 279 GH lines. The top-level architecture is a 4 by 26 array of MegaLABs. Communication between MegaLABs is accomplished by global H (horizontal) and V (vertical) wires, that switch at their intersection points. The H and V lines are segmented by a bidirectional segmentation buffer at the horizontal and vertical centers of the chip. In Fig. 3.10, We denote the use of a single (half-chip) H or V line as H or V and a double or full-chip line through the segmentation buffer as HH or VV. The 20K400 contains 100 H lines per MegaLAB row, and 80 V lines per LAB-column.

3.4 Proposed FPGA Interconnect Architectures

Design of large devices implies fundamental and efficient innovation in architecture to improve speed, density and software mapping time. Relying on industry experience with standard ASICs, we believe that partitioning and hierarchy becomes unavoidable for hardware and software developments. In fact most logic designs exhibit local connections, which implies a hierarchy in placement and routing of connections between logic blocks. The Tree-based interconnect architecture which takes advantage of this feature to provide smaller routing areas and more predictable timing behavior. Routability and interconnect area depend on switch boxes topology and signals bandwidth (in/out signals per cluster). In Tree-based interconnect topology, we use full cross bar switch boxes and we aim to exploit the flexibility to reduce signals bandwidth based on suitable partitioning approaches. In this section we describe a Tree-based interconnect model to improve logic density and speed of FPGAs.

3.4.1 Evolution of Tree-Based Interconnect Architecture

As illustrated in Fig. 3.11, in Tree-based interconnect architecture, Logic blocks and routing resources are partitioned into a multilevel clustered structure. Each cluster contains sub-clusters and a switch box allowing to connect external signals to sub-clusters. The previous studies regarding hierarchical topology described in Sect. 3.3.4, we consider the problem with different stand point. All Tree based networks presented in the Sect. 3.3.4 use bidirectional switches and wire tracks. This introduces considerable complication in both hardware network design and increases the load on routing tools [22, 23]. In Tree-based FPGA architecture we use only single-driver unidirectional wires. As proposed in [24], we build a fully hierarchical interconnect with inter-level signaling bandwidth, growing according to Rent's Rule. We consider only unidirectional signal wires. Logic Blocks represent the Tree leaves. Let N the number of LBs in the architecture. Gates are recursively partitioned into k equally sized sets at each level of the hierarchy. The principal interconnect occurs at each node of convergence in the hierarchy as presented in Fig. 3.11. At level ℓ in the hierarchy, each node has a fan-in from the lower level equal to $k * n_{out_{\ell-1}}$ signals and a fan-in from the upper level equal to n_{in_ℓ}. Similarly, it has a fan-out of $k * n_{in_{\ell-1}}$ toward the leaves and n_{out_ℓ} towards the root. At each level ℓ, we have n_{LB_ℓ} LBs, n_{in_ℓ} external inputs and n_{out_ℓ} external outputs. We are interested to evaluate wiring and switching growth. According to the architecture hierarchy and the Rent's rule growth we have:

$$N_{LB_\ell} = n^\ell$$

$$n_{in_\ell} = c_{in}.k^{\ell.p}$$

$$n_{out_\ell} = c_{out}.k^{\ell.p} \tag{3.3}$$

where c_{in} and c_{out} represent the number of leaf inputs and outputs and p is the Rent's growth factor.

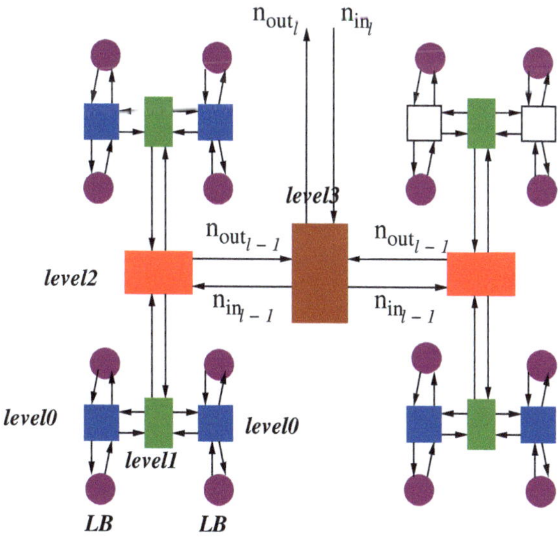

Fig. 3.11 TFPGA: Tree based FPGA interconnect

3.4.2 Wire Growth Model

First we consider how wiring resources grow in this structure. At each level ℓ of the hierarchy, each switching node has n_{in_ℓ} inputs and n_{out_ℓ} outputs. This makes the bisection width equal to $(c_{in} + c_{out})k^{\ell.p}$. Since $\forall \ell \in \{1, \ldots, \log_k(N)\}\ \ k^{\ell.p} \leq N$, the bisection width is $O(N^p)$. For a 2-dimensional network layout this bisection width must cross out of the subarray through the perimeter. Thus the perimeter of each subarray is $O(N^p)$. The area of the subarrays will be proportional to the square of its perimeter, making: $A_{subarray} \propto N^{2p}$. The area required for each Logic block (LB) based on wiring constraints, then goes as:

$$A_{wiring}(LB) \propto N^{2p-1}$$

3.4.3 Switch Growth Model

We have $k + 1$ distinct output directions from each node of convergence in the interconnect: k for the k leaves, plus one for the root. Allowing full connectivity within each Tree node, each of the k leaves picks its n_{in} inputs from the $(k-1) * n_{out}$ outputs from its siblings and from the n_{in} inputs from the parent node. The n_{out} outputs of this node are selected from the $k * n_{out}$ outputs from all k subtrees converging at this point. Each of the logical switching units is a fully-populated crossbar. At each level ℓ, the total switch number is:

$$N_{switch}(\ell) = [k * ((k-1)n_{out_{\ell-1}} + n_{in_\ell}) * n_{in_{\ell-1}}] + [(k * n_{out_{\ell-1}}) * n_{out_\ell})] \quad (3.4)$$

$$= [k^p(c_{in} + c_{out}) + (k-1)c_{out}]kc_{in}k^{2p(l-1)} \quad (3.5)$$

Looking across the number of LBs supported at level ℓ, we can count the number of switches per LB at each level:

$$N_{switch}(\ell) = \frac{[k^p(c_{in} + c_{out}) + (k-1)c_{out}]kc_{in}k^{2p(\ell-1)}}{k^\ell}$$

Summing across all levels we obtain:

$$N_{switch}(LB) = [k^p(c_{in} + c_{out}) + (k-1)c_{out}] \times c_{in} \sum_{\ell=1}^{\log_k(N)} k^{(2p-1)(\ell-1)} \quad (3.6)$$

$$N_{switch}(LB) = \begin{cases} O(1) & \text{if } p < 0.5 \\ O\left(\log_k(N)\right) & \text{ifp} = 0.5 \\ O\left(N^{2p-1}\right) & \text{ifp} > 0.5 \end{cases} \quad (3.7)$$

From Eq. 3.7 we have switching area per LB grows as $O(1)$, for $p < 0.5$, and $O\left(N^{2p-1}\right)$ for $p > 0.5$. The large area of switches relative to wires is one of the reasons that we will care about the number of switches required by the network. To reduce switches requirement we aim to reduce the interconnect Rent's parameter p. The architecture Rent's parameter is the minimum possible Rent's parameter of the architecture allowing all routability achievement of a given netlist. In the proposed architecture, we use fully populated switch boxes (crossbar). There is exactly one switch associated with each possible input to output connection, so routing is trivial and guaranteed. The architecture Rent's parameter corresponds exactly to the partitioned netlist Rent's parameter. In [25], authors showed that the resulting Rent's parameter is subject to the algorithm which generates the partitioning Tree. Thus in a Tree-based architecture switches requirement depends on the partitioning methodology. To determine the Rent's parameter of a netlist design we developed a multilevel partitioning methodology. In each hierarchical level, we determine the maximum number of inputs and outputs in all parts. Numbers of inputs and outputs of a cluster located at level ℓ of the Tree architecture are given by:

$$N_{in}(\ell) = \max_{part \in P(\ell)} N_{in}(part)$$

$$N_{out}(\ell) = \max_{part \in P(\ell)} N_{out}(part)$$

$P(\ell)$ is the set of parts in level ℓ.

3.5 Tree-Based Routing Interconnect

The Tree-based interconnect architecture uses *Butterfly Fat-Tree* (BFT) interconnect topology, where LBs (Logic Blocks) are grouped into clusters. Each cluster contains a switch block to connect local LBs. A switch block is divided into MSBs (Mini-switch Blocks). As illustrated in Fig. 3.12 The Tree-based interconnect architecture unifies two unidirectional programmable interconnect networks.

1. Downward Interconnect Network—The downward network is inspired from SPIN [26]. It is based on the Butterfly Fat-Tree (BFT) style interconnect [27] with linear populated and unidirectional switch boxes. Tree leaves correspond to logic blocks.
2. Upward Interconnect Network—The upward network connects the logic blocks outputs and the input Pads to the different levels of the Tree using hierarchy.

Each logic block contains one Look-Up-Table (with c_{in} inputs and $c_{out} = 1$ output), followed by a bypass Flip-Flop. The Logic Blocks (LBs) are grouped into k sized clusters and interconnect are organized into levels. Let $nb\ell$ denote the number of levels of a given architecture ($nb\ell = log_k(N)$). In each level ℓ we have $\frac{N}{k^\ell}$ clusters; C is the set of clusters in all levels. A cluster with index c belonging to level ℓ

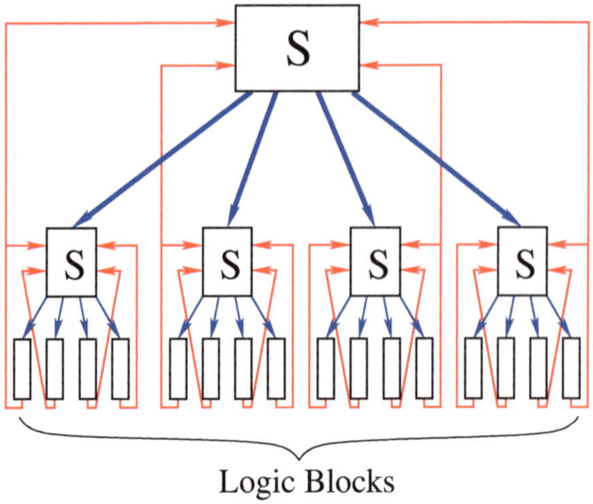

Fig. 3.12 Typical Tree-based interconnect architecture

is noted by $cluster(\ell, c)$. A cluster switch block is divided into separated Mini Switch Boxes (MSBs). Each MSB corresponds to a full crossbar. Each $cluster(\ell, c)$ where $\ell \geq 1$ contains a set of inputs $N_{in}(\ell)$, a set of outputs $N_{out}(\ell)$, a set of MSBs and k sub-clusters. Sub-clusters of $cluster(\ell, c)$ are $cluster(\ell - 1, k.c + i)$ where $i \in \{0, 1, 2, .., k-1\}$. k is called $cluster(\ell, c)$ arity. Let $nbMSB(\ell)$ the MSB number in a cluster in level ℓ. MSB with index m belonging to $cluster(\ell, c)$ is denoted $MSB(\ell, c, m)$. Each MSB contains k inputs driven by the upper level and 1 feedback coming from a leaf output pin. Each cluster in level 0 is denoted $cluster(0, c)$ or $leafcluster(c)$ and corresponds to the Logic Block (LB) and contains c_{in} inputs, 1 output, no MSBs and no sub-cluster. Each $cluster(\ell, c)$ where $\ell < nb\ell - 1$ has an owner in level ℓ' where $\ell' > \ell$ denoted $cluster(\ell', c \div k^{(\ell' - \ell)})$. We define for each $cluster(\ell, c)$ a position inside its owner in level $\ell + 1$ (direct owner) by the following function:

$$\begin{aligned} pos: \quad C &\longrightarrow \{0, 1, 2, .., k-1\} \\ cluster(\ell, c) &\longmapsto c \bmod k \end{aligned}$$

2 clusters belonging to level ℓ and with the same owner at level $\ell+1$ have 2 different positions. To get the cluster owner in level ℓ' of $cluster(\ell, c)$ ($\ell < \ell' \leq nb\ell - 1$) we define the function:

$$\begin{aligned} owner: C \times \mathbb{N} &\longrightarrow C \\ (cluster(\ell, c), \ell') &\longmapsto cluster(\ell', c \div k^{\ell' - \ell}) \end{aligned}$$

3.5.1 Tree-Based FPGA Architecture

As described in Sect. 3.5, in a Tree-based FPGA architecture [16, 17], the LBs are grouped into clusters and each cluster contains a switch block to connect local LBs. A switch block is divided into Mini switch Blocks (MSBs). The Tree-based FPGA architecture unifies two unidirectional upward and downward interconnection networks by using a *Butterfly-Fat-Tree* (BFT) topology to connect the Downward MSBs (DMSBs) and the Upward MSBs (UMSBs) to LBs inputs and outputs. As illustrated in Fig. 3.13, the UMSBs are used to allow LBs outputs to reach a large number of DMSBs and to reduce fanout on feedback lines. The UMSBs are organized in a way allowing LBs belonging to the same *owner-cluster* to reach exactly the same set of DMSBs at each level. Thus positions, inside the same cluster, are equivalent, and LBs can negotiate with their siblings about the use of a larger number of DMSBs depending on their fanout sizes. For example in Fig. 3.13, an LB ouput can reach all 4 DMSBs of its owner cluster at level 1 and all the 16 DMSBs of its owner cluster at level 2. Therefore the external I/O pads, clusters or logic-block positions inside the direct owner cluster become equivalent and there is no need to re-arrange them anymore. The input and output pads are grouped into specific clusters and are connected to UMSBs and DMSBs, respectively as presented in Fig. 3.13. Thus, all input and output pads can reach any LBs of the architecture. As presented in Fig. 3.13, the programmable interconnects of a Tree-based FPGA architecture are arranged in a multilevel network with the switch blocks placed at different tree levels using a *Butterfly-Fat-Tree* network topology.

Using UMSBs and DMSBs greatly enhances routability of the Tree-based FPGA architecture, but it increases the total number of switch requirements. However this increase can be compensated by depopulating in/out signals bandwidth of clusters at every level. In fact, netlists implemented on FPGA architecture often

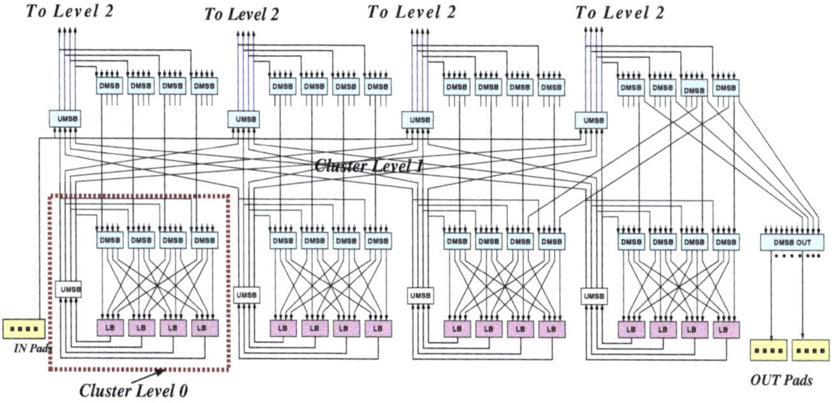

Fig. 3.13 An illustration of two-level Tree-based FPGA interconnect structure using two unidirectional upward and downward interconnects

communicate locally (intra-clusters) and this fact can be exploited to reduce the bandwidth of signals with inter-clusters communication. A good estimation of netlists communication locality is given by Rent's Rule [28]. Based on this estimation authors in [29] showed that most netlist Rent's parameters range between 0.5 and 0.65. As shown in Fig. 3.13, the number of DMSBs of a cluster located at level ℓ is equal to the number of inputs of a cluster located at level $\ell - 1$. The UMSBs (Upward MSBs) allow LBs outputs to reach a large number of DMSBs and to reduce fanout on feedback lines. The number of UMSBs of a cluster located at level ℓ is equal to the number of outputs of a cluster located at level $\ell - 1$. The Table 3.2 shows the switch growth of Tree-based interconnect architecture with seven tree levels. It

Table 3.2 Switch growth model of Tree-based interconnect

Tree Levels = 7	Arity = 4, Cluster size = 4 Arch = 4 × 4 × 4 × 4 × 4 × 4 × 4					
Levels	LUT3	LUT4	LUT5	LUT6	LUT7	LUT8
0	4	5	6	7	8	9
1	16	20	24	28	32	36
2	64	80	96	112	128	144
3	256	320	384	448	512	576
4	1024	1280	1536	1792	2048	2304
5	4096	5120	6144	7168	8192	9216
6	16,384	20,480	24,576	28,672	32,768	36,864
Tree Levels = 7	Arity = 4, Cluster size = 5 Arch = 5 × 5 × 5 × 5 × 5 × 5 × 5					
Levels	LUT3	LUT4	LUT5	LUT6	LUT7	LUT8
0	4	5	6	7	8	9
1	20	25	30	35	40	45
2	100	125	150	175	200	225
3	500	625	750	875	1000	1125
4	2500	3125	3750	4375	5000	5625
5	12,500	15,625	18,750	21,875	25,000	28,125
6	62,500	78,125	93,750	109,375	125,000	140,625
Tree Levels = 7	Arity = 4, Cluster size = 6 Arch = 6 × 6 × 6 × 6 × 6 × 6 × 6					
Levels	LUT3	LUT4	LUT5	LUT6	LUT7	LUT8
0	4	5	6	7	8	9
1	24	30	36	42	48	54
2	144	180	216	252	288	324
3	864	900	1296	1512	1728	1944
4	5184	5400	7776	9072	10,368	11,664
5	31,104	32,400	46,656	54,432	62,208	69,984
6	186,624	194,400	279,936	326,592	373,248	419,904

also shows, the switch growth rate is slower in the initial levels of Tree, however it increases exponentially as the Tree grows to higher levels and this trend is also same for the wire length at higher levels.

The Table 3.2 the impact of LUT and cluster size on total number of switch requirements. The UMSBs are organized in a way allowing LBs belonging to the same *owner cluster* to reach exactly the same set of DMSBs at each level. Thus positions, inside the same cluster, are equivalent, and LBs can negotiate with their siblings the use of a larger number of DMSBs depending on their fanout. As shown in Fig. 3.13, input and output pads are grouped into specific clusters and are connected to UMSBs and DMSBs, respectively. Thus, input pads can reach all LBs of the architecture, and can also be reached by different paths. The cluster size and the level where it is connected can be modified to obtain the best design fit. Similarly the output pads are connected to all DMSBs of the upper level and they can be reached from all LBs through different paths. The main focus of this book is to describe the development of tools and technologies for the implementation 3D Tree-based FPGA architectures. Nevertheless this book also provide details about the other varients of Tree-based FPGA architectures. One such varient is described next section.

3.6 Unified Mesh- and Tree-Based Interconnect

The Tree-based architecture is better optimized in terms of switches number than the island-style Mesh-based architecture. Nevertheless the Butterfly Fat-Tree topology, in stand alone mode, is very penalizing in terms of physical layout generation. As illustrated in Fig. 3.14, it does not support scalability and does not fit with a planar chip structure especially for large circuits unless some dramatic changes happen in the interconnect layout, which is the aim of this work. In the lower levels of the

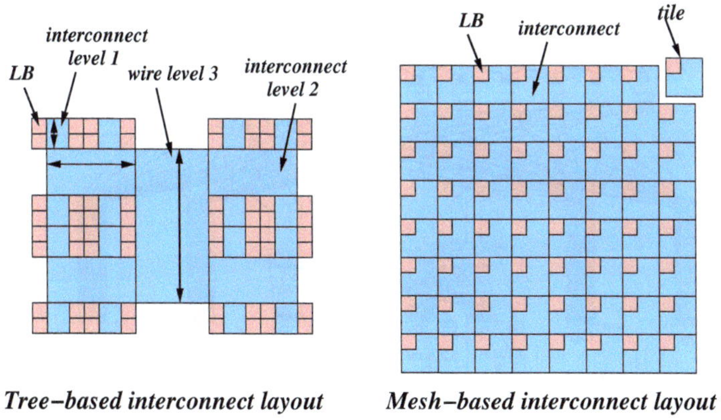

Fig. 3.14 Node of Mesh of Tree architecture

Tree-based interconnect, clusters are small and close by, but as we get higher into the tree (connecting large clusters), wiring distances increase. Interconnect in the higher part of the tree may need to be subdivided. Our idea is to use Tree topology as an intra-cluster interconnect and to use Mesh topology to achieve inter-clusters interconnection. In this way we can control the tree size and generate a layout with any size by tiles abutment.

The Mesh-of-Tree interconnect topology (MoT) is built as a matrix of abutted nodes presented in Fig. 3.15. Each node has a Tree-based intra-cluster interconnect. The resulting network corresponds to a Mesh of clusters (each one encapsulating the intra-cluster interconnect and the LBs). Clusters surrounded by Mesh interconnect are called Mesh clusters and clusters included in the different levels of the tree are called tree clusters. This topology is proposed as an alternative to the common cluster-based Mesh architectures. As shown in Fig. 3.15, there exist different ways to connect signals to the LUT input muxes. In Xilinx Virtex architectures [30], the routing tracks are connected directly to the input muxes. In the VPR architecture [6] and the Altera Stratix architecture [31], the routing tracks are connected to the input muxes via an intermediate level of muxes called connection block. The VPR-style interconnect has a sparsely populated connection block and a fully populated intra-cluster crossbar. The fully populated intra-cluster crossbar is simple but takes no advantage of the logical equivalence of LUT inputs and induces a significant inefficiency. An improved Mesh-based island style interconnect presented in [22]. They proposed an approach to generate highly routable sparse connection block. Furthermore, they showed that the intra-cluster full crossbar can be depopulated to achieve

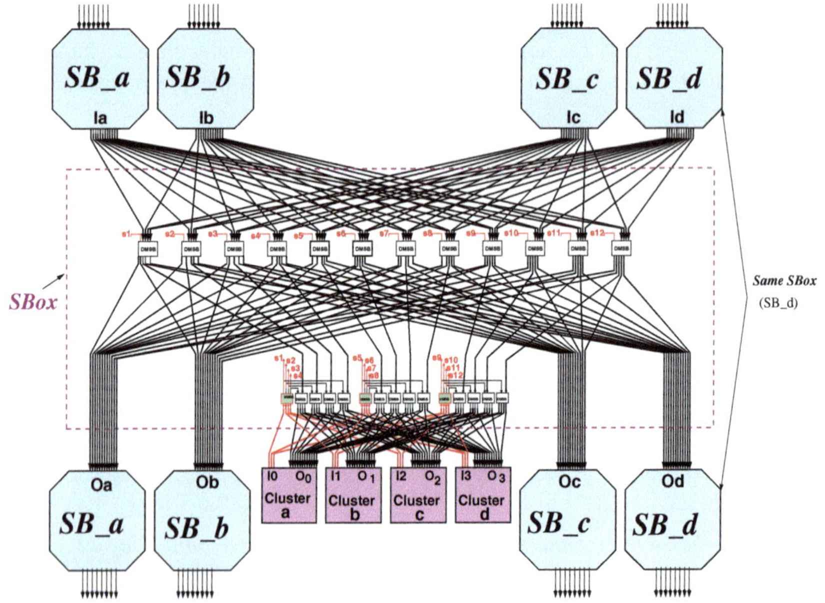

Fig. 3.15 Mesh of Tree SBOX architecture

significant area reduction without performance degradation. A practical example is Stratix, which depopulates this crossbar by 50 % [31]. All studies considered the connection block interconnect level and the intra-cluster crossbar separately. In [32], authors investigate joint optimization of both crossbars and proposed a new class of efficient topology. Nevertheless in the intra-cluster crossbar they optimized only the part connecting external signals to LBs inputs. Using a full crossbar to connect feedbacks (LBs outputs) to LBs inputs is very penalizing and imposes a very low bound on the cluster LBs number. For example we assume we have a cluster with 256 LBs and we use a full crossbar to connect feedbacks to 4-Lut inputs. This means we need 262×10^3 switches to route clusters internal signals only, which is very expensive. In the proposed Mesh of Tree architecture, our first contribution corresponds to a joint optimization of connection blocks and inta-cluster interconnect topologies. We optimize both crossbars: (1) connecting external signals to LBs inputs (2) connecting feedbacks to LBs inputs. Our second contribution consists in using only single-driver interconnect based on unidirectional wires. As illustrated in [23, 33], single-driver interconnect has a good impact on density improvement.

3.6.1 Cluster Local Interconnect

Mesh clusters are composed of Logic Blocks (LBs) which communicate within a programmable local interconnect. The intra-cluster interconnect is organized as a Tree and has the topology similar to Tree-based FPGA. Mesh Clusters input and output pins are connected to LBs as MFPGA input and output pads. Figures 3.15 and 3.16 illustrates Mesh cluster Tree-based local networks and its interface with the Mesh-based interconnect. As illustrated in Figs. 3.15 and 3.17, each cluster is connected to the 4 adjacent channel tracks. The cluster input and output connectors are equally distributed on the 4 sides. In all sides we have the same number of inputs and outputs. As shown in Fig. 3.17, input and output signals are grouped into clusters situated at level ℓ of the Tree:

- Each cluster of input signals contains 4 inputs connected to the 4 adjacent channels. As shown in Fig. 3.17, each input is connected to all UMSBs located at level $\ell + 1$ of the Tree. In this way the 4 inputs are logically equivalent and can reach all Tree LBs.
- Each cluster of output signals contains 4 outputs connected to the 4 adjacent switch boxes. As shown in Fig. 3.17, each output is connected to all DMSBs placed at level $\ell + 1$ of the Tree. In this way the 4 outputs are logically equivalent and can be reached from all the Tree LBs.

This distribution has an important impact on routability and eliminates constraints in the placement of Logic Blocks inside clusters. All four Mesh cluster sides have the same number of inputs and outputs. Side inputs and outputs numbers depend on the number of Tree leaves and on the level of UMSBs and DMSBs where they are connected.

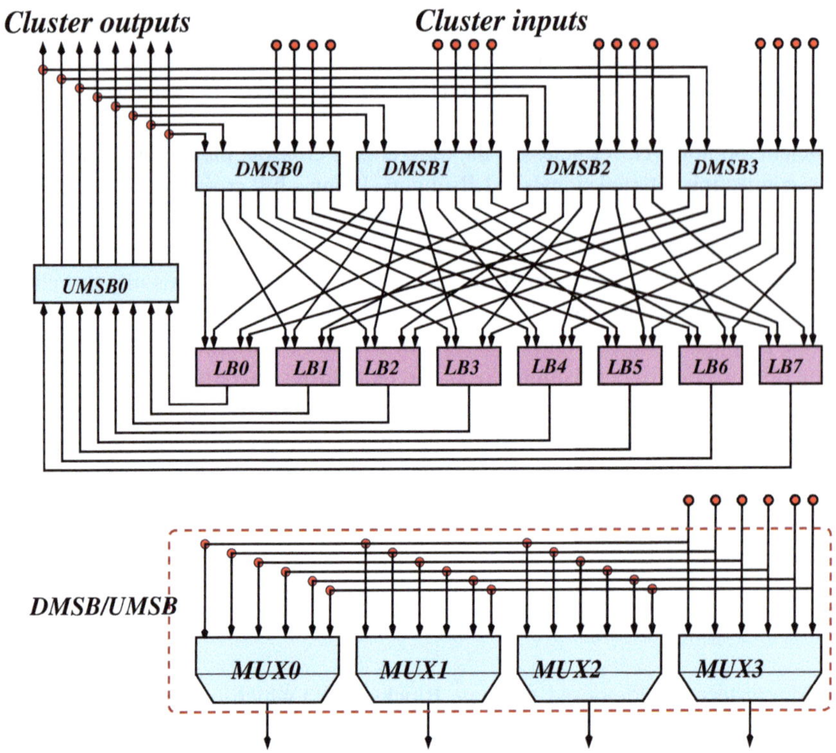

Fig. 3.16 Mesh of Tree FPGA cluster architecture

$$Nb_{in} = \frac{N}{k^{\ell_{in}+1}} \tag{3.8}$$

$$Nb_{out} = \frac{N}{k^{\ell_{out}+1}} \tag{3.9}$$

k is Tree clusters arity. N is the number of Tree leaves. ℓ_{in} and ℓ_{out} are respectively levels where input and output signals are connected

3.6.2 Mesh-Based Routing Interconnect

In the Mesh interconnect we use only single-driver unidirectional wires, in fact in [23], authors show that single-driver based interconnect leads to a 25 % improvement in area density. Each Mesh cluster is surrounded by 4 channels which are connected by Switch Boxes (SB). We do not use connection blocks in the Mesh to connect channel tracks to cluster inputs and outputs. In fact, as presented in [32],

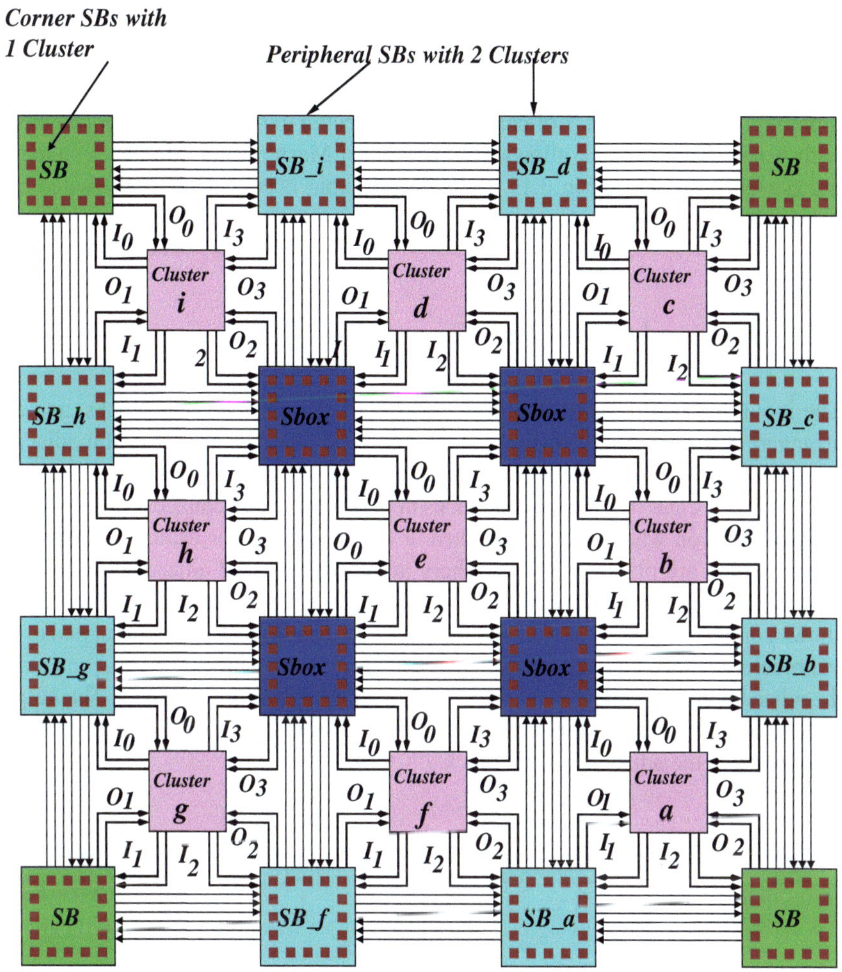

Fig. 3.17 Island style representation of Mesh-of-Tree (MoT)-based FPGA architecture

interconnect is better optimized when the connection block is combined with the cluster local interconnect. As described in Fig. 3.17, Mesh cluster input signals are connected to the 4 adjacent channels tracks. Thus, channel width W is given by:

$$W = \frac{Nb_{in}}{4} = \frac{N}{k^{\ell_{in}+1}} \tag{3.10}$$

Consequently, the channel width depends on the cluster inputs number and it is very expensive to modify it in terms of routing resources. In fact modifying the channel width induces modification of the cluster interface and consequently the Tree interconnect structure. A Mesh Switch Box (SB) allows to connect horizontal and vertical channel tracks together and also to clusters outputs. SB inputs come from the 4 channel tracks and the 4 adjacent clusters outputs. Since we use a single-driver based interconnect, each SB output is driven by a multiplexer. SB has a disjoint topology. The input track j of channel i is connected to output track j of channel h with $h \neq i$. Switch block allows also to connect Mesh cluster outputs to channels tracks. As illustrated in Fig. 3.17 each cluster output is connected to all switch box outputs located at the 4 sides.

3.6.3 Input and Output Pads Connection

The input and output pads of MoT-based FPGA are grouped into blocks, and are arranged at the periphery of the architecture. They are connected to the adjacent Switch blocks located at the periphery of the chip as illustrated in Fig. 3.18. From the Fig. 3.17, it is clear that the switch blocks at periphery either connects to one cluster or two based on where they are located at the periphery. For example, the switch blocks which are placed at the periphery are connected to 2 adjacent clusters and switch blocks placed at corners of FPGA connects only to one cluster. Figure 3.18 shows connections between pad and the adjacent switch block. Each input pad of an I/O block is connected to all UMSBs of the adjacent switch block, and thus can reach the adjacent clusters and can be connected to adjacent routing channels. The output pads of an I/O block are grouped into specific cluster and connected to all DMSBs of level $\ell 1$ of the adjacent switch block.

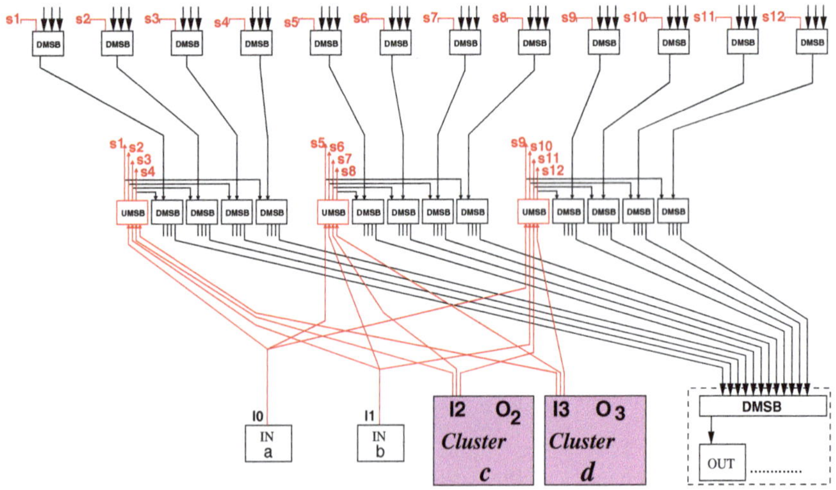

Fig. 3.18 Peripheral Switch block of MoT-based FPGA architecture: connection of Peripheral Switch block with In/Out Pads

3.7 Summary

The interconnect structure of a Mesh-based FPGA is generally designed to maximize logic utilization. Hierarchical FPGAs belong to the class of FPGA architectures thats are designed to increase the interconnect utilization at the expense of logic utilization. The philosophy behind hierarchical architectures is increased silicon utilization through efficient use of the interconnect structure (which may account for 80–90 % of the total area of Mesh-based FPGAs). However we proposed two different FPGA architectures with modified interconnect architecture in this chapter. The Tree-based FPGA architecture uses two unidirectional interconnect network to connect the inputs and outputs of LBs to SBs. Using the Tree-based FPGA architecture, we achieved 56 % reduction in number of switched used compared to Mesh-based FPGA architectures. The main issues with Tree-based interconnect structure is the path delay increases exponentially as Tree grows to higher levels. We discuses the solutions to reduce interconnect delay using 3D Technology in Chap. 6. Another FPGA architecture proposed in this chapter is MoT-based FPGA. MoT-based FPGA has an unified Mesh and Tree architecture to combine the advantages of both in one platform. MoT-based FPGA is an interesting architecture in terms of it flexibility. We can used this architecture to design and implement 2.5D multi-FPGAs and 3D multi-tier FPGAs, however the focus of this book is on design methodologies of 3D Tree-based FPGAs.

References

1. W.S. Carter, K. Duong, R.H. Freeman, H.C. Hsieh, J.Y. Ja, E. Mahoney, L.T. Ngo, S.L. Sze, A user programmable reconfigurable logic array, in *Proceedings of IEEE, Custom Integrated Circuits Conference*, May 1986, pp. 233-235
2. J. Rose, S. Brown, Flexibility of interconnect structures for field programmable gate arrays. IEEE J. Solid State Circuits **26**, 277–282 (1991)
3. S.D. Brown, R.J. Francis, J. Rose, Z.G. Vranesic, *Field Programmable Gate Arrays* (Kulwer, Norwell, 1992)
4. S. Trimberger et al., A re-programmable gate array and applications. Proc. IEEE **81**, 1030–1041 (1993)
5. V. Betz, J. Rose, How much logic should go in an FPGA Logic Block?. IEEE Des. Test Comput. **15**(1), 10–15 (1998)
6. V. Betz J. Rose, A. Marquardt, *Architecture and CAD for Deep Sub-micron FPGAs* (Kluwer, Norwell, 1999)
7. Xilinx Inc., Two flows for partial reconfiguration: module based or difference based, 2004. http://www.xilinx.com/bvdocs/appnotes/xapp290.pdf
8. E. Ahmed, J. Rose, The effect of LUT and cluster size on deep-sub-micron FPGA performance and density. IEEE Trans. Very Large Scale Integr. (VLSI) Syst. **22**(3), 288–298 (2004)
9. G. Lemieux, D. Lewis, Analytical framework for switch block design, in *International Conference on Field Programmable Logic (FPL)*, pp. 122–131, 2002
10. A.A. Aggarwal, D.M. Lewis, Routing architectures for hierarchical field programmable gate arrays, in *Proceedings of the 1994 IEEE International Conference on Computer Design: VLSI in Computer & Processors*, 10–12 Oct 1994, pp. 475–478

11. V.C. Chan, D.M. Lewis, Area-speed tradeoffs for hierarchical field-programmable gate arrays, in *Proceedings of the 1996 ACM Fourth International Symposium on Field-Programmable Gate Arrays*, 11–13 Feb 1996, Monterey, California, USA, pp. 51–57
12. Yen-Tai Lai, Ping-Tsung Wang, Hierarchical interconnection structures for field programmable gate arrays. IEEE Trans. Very Large Scale Integr. (VLSI) Syst. **5**(2), 186–196 (1997)
13. A. DeHon, Balancing interconnect and computation in a reconfigurable computing array (or, why you don't really want 100% LUT utilization), in *Proceedings of the 1999 ACM/SIGDA Seventh International Symposium on Field Programmable Gate Arrays*, 21–23 Feb 1999, Monterey, California, USA, pp. 69–78
14. W. Tsu, K. Macy, A. Joshi, R. Huang, N. Walker, T. Tung, O. Rowhani, V. George, J. Wawrzynek, A. DeHon, HSRA: high-speed, hierarchical synchronous reconfigurable array, in *Proceedings of the 1999 ACM/SIGDA Seventh International Symposium on Field Programmable Gate Arrays*, 21–23 Feb 1999, Monterey, California, USA, pp. 125–134
15. Andre DeHon, Unifying mesh- and tree-based programmable interconnect. IEEE Trans. Very Large Scale Integr. (VLSI) Syst. **12**(10), 1051–1065 (2004)
16. Z. Marrakchi, H. Mrabet, C. Masson, H. Mehrez, Mesh of tree: unifying Mesh and MFPGA for better device performances, in *NOCS 2007*, pp. 243–252, 2007
17. Z. Marrakchi, H. Mrabet, U. Farooq, H. Mehrez, FPGA interconnect topologies exploration. Int. J. Reconfig. Comput. **2009** (2009)
18. Andre DeHon, Raphael Rubin, Design of FPGA interconnect for multilevel metallization. IEEE Trans. Very Large Scale Integr. (VLSI) Syst. **12**(10), 1038–1050 (2004)
19. J. Rose, R. Francis, D. Lewis, P. Chow, Architecture of field-programmable gate arrays: the effect of logic functionality on area efficiency. IEEE J. Solid State Circuits **25**, 1217–1225 (1990)
20. V. Adler, E.G. Firedman, Repeater insertion to reduce delay and power in RC tree structures, in *Conference on Signals, Systems and Computers*, pp. 749–752, 1997
21. M. Hutton, K. Adibasamii, A. Leaver, Timing driven placement for hierarchical programmable logic devices, in *International Symposium on Field Programmable Gate Arrays*, pp. 3–11, 2001
22. G. Lemieux, D. Lewis, *Design of Interconnection Networks for Programmable Logic* (Springer, formerly Kluwer Academic Publishers, Norwell, 2004). ISBN: 1-4020-7700-9
23. G. Lemieux, E.Lee, M.Tom, A.Yu, Directional and single-driver wires in FPGA interconnect, in *IEEE International Conference on Field-Programmable Technology, FPT-2004*, pp. 41–48, 2004
24. A. DeHon, Reconfigurable architectures for general-purpose computing. Ph.D. dissertation, Department of Electrical Engineering and Computer Science, Massachusetts Institute of Technology, 1996
25. L. Hagen, A.B. Kahng, F.J. Kurdahi, C. Ramachandran, On the intrinsic rent parameter and spectra-based partitioning methodologies. IEEE Trans. Comput. Aided Des. **13**(1), 27–37 (1994)
26. P. Guerrier, A. Greiner, A generic architecture for on chip packet-switched interconnections, in *Proceedings of the Design Automated and Test in Europe Conference (DATE)*, Paris, France pp. 250–256, 2000
27. C. Leiserson et al., Fat-trees: universal networks for hardware efficient supercomputing. IEEE Trans. Comput. **C-34**(10), 892–901 (1985)
28. B. Landman, R. Russo, On a pin versus block relationship for partitions of logic graphs. IEEE Trans. Comput. **20**(12), 1469–1479 (1971)
29. J. Pistorius, M. Hutton, Placement rent exponent calculation methods, temporal behaviour and FPGA architecture evaluation, in *Proceedings of the International Workshop on System Level Interconnect Prediction*, Apr 2003, Monterey, Calif, USA, pp. 31–38
30. Vertex-5, Xilinx Inc., Vertex-5: multi-platform FPGA. http://www.xilinx.com/products/silicon_solutions/fpga/vertex/vertex5/
31. D.Lewis et al., The stratix logic and routing architecture, in *International Symposium on Field Programmable Gate Arrays, FPGA-2003*, Feb 2003, pp. 12–20

32. W. Feng, S. Kaptanoglu, Designing efficient input interconnect blocks for LUT clusters using counting and entropy, in *International Symposium on Field Programmble Gate Array, (FPGA-2007)*, pp. 23–32, 2007
33. G. Lemieux, D. Lewis, Using sparse crossbars with LUT clusters, in *Proceedings of ACM/SIGDA International Symposium on FPGAs*, Feb 2001, pp. 59–68

Chapter 4
Two Dimensional FPGAs: Configuration and CAD Flow

Abstract FPGA architectures have been intensely investigated over the past two decades. A major aspect of FPGA architecture research is the development of Computer Aided Design (CAD) tools for design and implementation of fast and high density FPGAs and mapping applications to it. It is well established that the quality of an FPGA based implementation is largely determined by the effectiveness of accompanying suite of CAD tools. Benefits of an otherwise well designed, feature rich FPGA architecture might be impaired if the CAD tools cannot take advantage of the features that the modern FPGA design provides. Thus, CAD algorithm research is essential to the necessary architectural advancement to narrow down the performance gaps between FPGAs and other computational devices like ASICs. This chapter discuss different algorithms and methodologies used to create 2D FPGA placement, routing, mapping application etc.

4.1 Introduction

FPGA architectures have been intensely investigated over the past two decades. A major aspect of FPGA architecture research is the development of Computer Aided Design (CAD) tools for design and implementation of fast and high density FPGAs and mapping applications to it. It is well established that the quality of an FPGA-based implementation is largely determined by the effectiveness of accompanying suite of CAD tools. Benefits of an otherwise well designed, feature rich FPGA architecture might be impaired if the CAD tools cannot take advantage of the features that the modern FPGA design provides. Thus, CAD algorithm research is essential to the necessary architectural advancement to narrow the performance gaps between FPGAs and other computational devices like ASICs. The process of converting a circuit description into a format that can be loaded into an FPGA can be roughly divided into five distinct steps, namely synthesis, technology mapping, clustering, placement, and routing. The final output of FPGA CAD tools is a bitstream that configures the state of memory bits in FPGA. The contents of configuration memory determines the logical function implemented. Figure 4.1 presented the CAD flow developed for the design and architecture exploration of 3D Tree-based FPGA.

V. Pangracious et al., *Three-Dimensional Design Methodologies for Tree-based FPGA Architecture*, Lecture Notes in Electrical Engineering 350, DOI 10.1007/978-3-319-19174-4_4

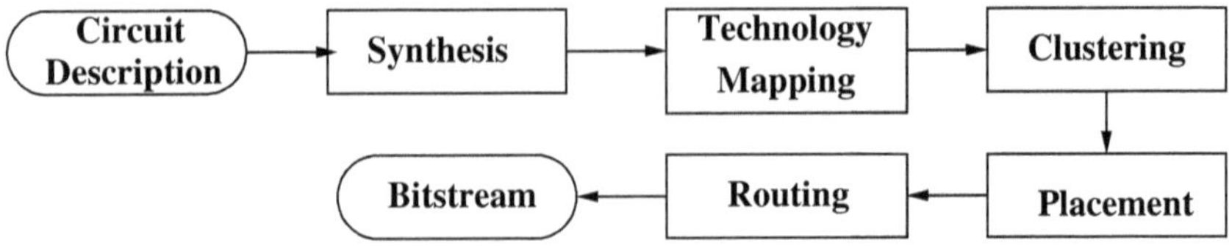

Fig. 4.1 3D Tree-based FPGA place and route

4.2 Circuit Synthesis

Synthesis involves translating a circuit description, traditionally written in a hardware description language (HDL) (e.g. VHDL or Verilog), into a gate-level representation. The gate-level representation is a network consisting of Boolean logic gates and flipflops. A logic optimization process is used to remove the redundant logic from the netlist and simplify the logic whenever possible. There are no FPGA-specific optimizations performed during synthesis since this is normally a technology independent step. As presented in Fig. 4.1, the operation is independent of the architecture. In our flow we use SIS [1] synthesis tool. SIS requires architecture parameters like k, the LUT input number. As presented in Fig. 4.1, this tool depends only on LUT size and can target any interconnect topology.

4.3 Technology Mapping

The output from synthesis tools is a circuit description of Boolean logic gates, flipflops and wiring connections between these elements. The circuit can also be represented by a Directed Acyclic Graph (DAG). Each node in the graph represents a gate, flip-flop, primary input or primary output. Each edge in the graph represents a connection between two circuit elements. Figure 4.2 shows an example of a DAG representation of a circuit. Given a library of cells, the technology mapping problem can be expressed as finding a network of cells that implements the Boolean network. In the FPGA technology mapping problem, the library of cells is composed of k-input LUTs and flip-flops. Therefore, FPGA technology mapping involves transforming the Boolean network into k-bounded cells. Each cell can then be implemented as an independent k-LUT. Figure 4.3 shows an example of transforming a Boolean network into k-bounded cells. Technology mapping algorithms can optimize a design for a set of objectives including depth, area or power. The FlowMap algorithm [2] is the most widely used academic tool for FPGA technology mapping. FlowMap is a breakthrough in FPGA technology mapping because it is able to find a depth-optimal solution in polynomial time. FlowMap guarantees depth optimality at the expense of logic duplication. Since the introduction of FlowMap, numerous technology mappers

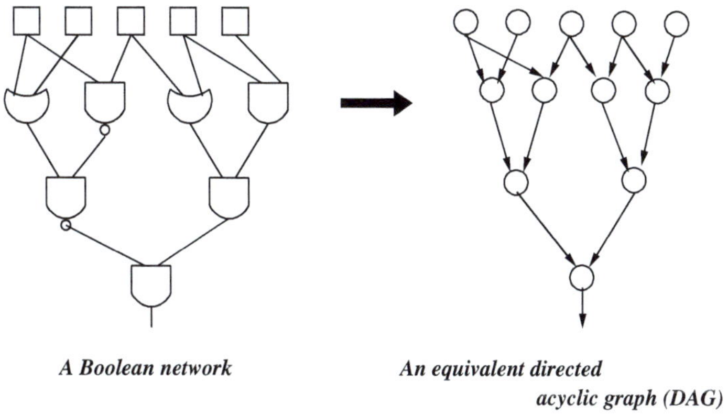

Fig. 4.2 Directed acyclic graph representation of a circuit

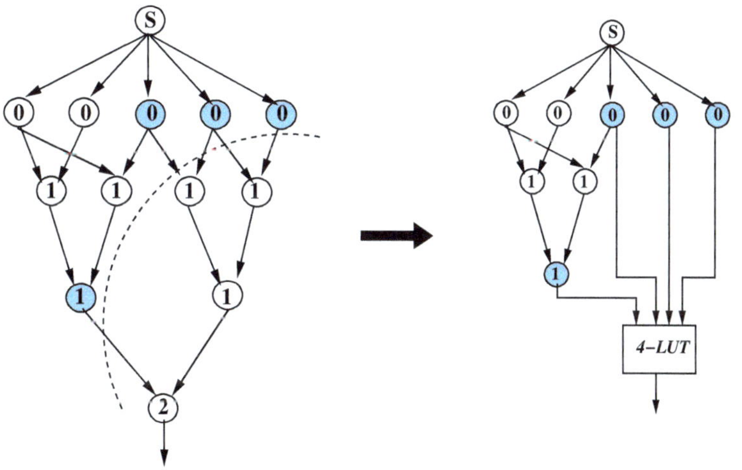

Fig. 4.3 Example of technology mapping

have been designed that optimize for area and run-time while still maintaining the depth-optimality of the circuit [3–5]. The result of the technology mapping step generates a network of k-bounded LUTs and flip-flops.

4.4 Clustering

The logic elements in a Mesh-based FPGA are typically arranged in two levels of hierarchy. The first level consists of logic blocks (LBs) which are *k-input* LUT and flip-flop pairs. The second level hierarchy groups *k* LBs together to form logic blocks clusters. The clustering phase of the FPGA CAD flow is the process of forming groups of k LBs. These clusters can then be mapped directly to a logic element on an FPGA. Figure 4.4 shows an example of the clustering process. Clustering algorithms can be broadly categorized into three general approaches, namely top-down [6, 7], depth-optimal [8, 9] and bottom-up [10–12]. Top-down approaches partition the LBs into clusters by successively subdividing the network or by iteratively moving LBs between parts. Depth-optimal solutions attempt to minimize delay at the expense of logic duplication. Bottom-up approaches are generally preferred for FPGA CAD tools due to their fast run times and reasonable timing delays. They only consider local connectivity information and can easily satisfy clusters pin constraints. Top-down approaches offer the best solutions; however, their computational complexity can be prohibitive.

Algorithm 1: Pseudo code of the V-Pack Algorithm

```
Data: Pseudo code of the V-Pack Algorithm [13]
Result: packed LBs
UnclusteredLBs = Pattern-Match-To-LBs(LUTs,Registers);
LogicClusters = NULL;
while UnclusteredLBs != NULL do
    C = GetLBwithMostUsedInputs(UnclusteredLBs);
    while | C |< k do
        /*cluster is not full*/
        BestLB = MaxAttractionLegalLB(C,UnclusteredLBs);
        if BestLB == NULL then
            /*No LB can be added to this cluster*/
            break;
        end
        UnclusteredLBs = UnclusteredLB − BestLB;
        C = C ∪ BestLB;
    end
    if | C |< k then
        /*Cluster is not full – try hill climbing*/
        while | C |< k do
            BestLB = MinClusterInputIncreaseLB(C,UnclusteredLBs);
            C = C ∪ BestLB;
            UnclusteredLBs = UnclusteredLB − BestLB;
        end
        if ClusterIsIllegal(C) then
            RestoreToLastLegalState(C,UnclusteredLBs);
        end
    end
    LogicClusters = LogicClusters ∪ C;
end
```

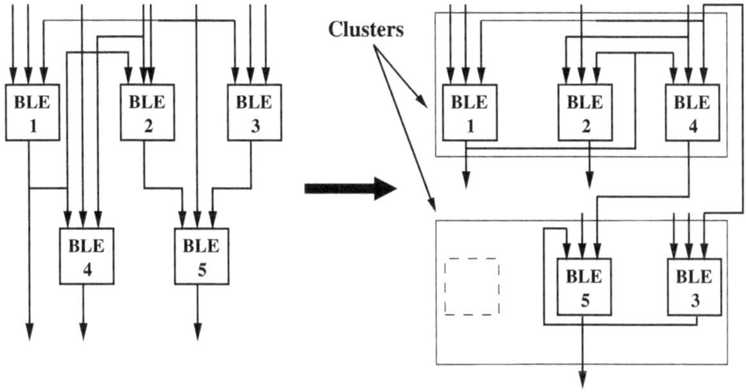

Fig. 4.4 Example of clustering

4.4.1 Bottom-Up Approaches

Bottom-up approaches build clusters sequentially one at a time. The process starts by choosing an LB which acts as a cluster seed. LBs are then greedily selected and added to the cluster, applying various attraction functions. The VPack [10] attraction function is based on the number of shared nets between a candidate LB and the LBs that are already in the cluster. For each cluster, the attraction function is used to select a seed LB from the set of all LBs that have not already been packed. After packing a seed LB into the new cluster, a second attraction function selects new LBs to pack into the cluster. LBs are packed into the cluster until the cluster reaches full capacity or all cluster inputs have been used. If all cluster inputs become occupied before this cluster reaches full capacity, a hill-climbing technique is applied, searching for LBs that do not increase the number of inputs used by the cluster. The VPack pseudo-code is outlined in Algorithm 1. T-VPack [13] is a timing driven version of VPack which gives added weight to grouping LBs on the critical path together. The algorithm is identical to VPack, however, the attraction functions which select the LBs to be packed into the clusters are different. The VPack seed function chooses LBs with the most used inputs, whereas the T-VPack seed function chooses LBs that are on the most critical path. VPack's second attraction function chooses LBs with the largest number of connections with the LBs already packed into the cluster. T-VPack's second attraction function has two components for a LB B being considered for cluster C:

$$Attraction(B, C) = \alpha.Crit(B) + (1 - \alpha)\frac{\mid Nets(B) \cap Nets(C) \mid}{G} \tag{4.1}$$

where $Crit(B)$ is a measure of how close LB B is to being on the critical path, $Nets(B)$ is the set of nets connected to LB B, $Nets(C)$ is the set of nets connected to the LBs already selected for cluster C, α is a user-defined constant which determines the relative importance of the attraction components, and G is a normalizing factor. The first component of T-VPack's second attraction function chooses critical-path LBs, and the second chooses LBs that share many connections with the LBs already packed into the cluster. By initializing and then packing clusters with critical-path LBs, the algorithm is able to absorb long sequences of critical-path LBs into clusters. This minimizes circuit delay since the local interconnect within the cluster is significantly faster than the global interconnect of the FPGA. RPack [12] improves routability of a circuit by introducing a new set of routability metrics. RPack significantly reduced the required channel widths required by circuits compared to VPack. T-RPack [12] is a timing driven version of RPack which is similar to T-VPack by giving added weight to grouping LBs on the critical path. iRAC [11] improves the routability of circuits even further by using an attraction function that attempts to encapsulate as many low fanout nets as possible within a cluster. If a net can be completely encapsulated within a cluster, there is no need to route that net in the external routing network. By encapsulating as many nets as possible within clusters, routability is improved because there are less external nets to route in total.

4.4.2 Top-Down Approaches

The K-way partitioning problem seeks to minimize a given cost function of such an assignment. A standard cost function is net cut, which is the number of hyperedges that span more than one partition, or more generally, the sum of weights of such edges. Constraints are typically imposed on the solution, and make the problem difficult. For example some vertices can be fixed in their parts or the total vertex weight in each part partition must be limited (balance constraint and FPGA clusters size). With balance constraints, the problem of partitioning optimally a hypergraph is known to be NP-hard [14, 15]. However, since partitioning is critical in several practical applications, heuristic algorithms were developed with near-linear runtime. Such move-based heuristics for k-way hypergraph partitioning appear in [16–18].

4.4.2.1 Fiduccia-Mattheyses Algorithm

The Fiduccia-Mattheyses (FM) heuristics works by prioritizing moves by gain. A move changes to which partition a particular vertex belongs, and the gain is the corresponding change of the cost function. After each vertex is moved, gains for connected modules are updated.

Algorithm 2: Pseudo-code for FM heurisctic [19]

```
partitioning = initial_solution;
while solution quality improves do
    Initialize gain_container from partitioning;
    solution_cost = partitioning.get_cost();
    while not all vertices locked do
        move = choose_move();
        solution_cost += gain_container.get_gain(move);
        gain_container.lock_vertex(move.vertex());
        gain_update(move);
        partitioning.apply(move);
    end
    roll back partitioning to best seen solution;
    gain_container.unlock_all();
end
```

The *Fiduccia-Mattheyses* or (FM) heuristic for partitioning hypergraphs [17] is an iterative improvement algorithm. FM starts with a possibly random solution and changes the solution by a sequence of moves which are organized as passes. At the beginning of a pass, all vertices are free to move (unlocked), and each possible move is labeled with the immediate change to the cost it would cause; this is called the gain of the move (positive gains reduce solution cost, while negative gains increase it). Iteratively, a move with highest gain is selected and executed, and the moving vertex is locked, i.e., is not allowed to move again during that pass. Since moving a vertex can change gains of adjacent vertices, after a move is executed all affected gains are updated. Selection and execution of a best-gain move, followed by gain update, are repeated until every vertex is locked. Then, the best solution seen during the pass is adopted as the starting solution of the next pass. The algorithm terminates when a pass fails to improve solution quality. Pseudo-code for the FM heuristic is given in Algorithm 2. The FM algorithm has 3 main components (1) computation of initial gain values at the beginning of a pass; (2) the retrieval of the best-gain (feasible) move; and (3) the update of all affected gain values after a move is made. One contribution of Fiduccia and Mattheyses lies in observing that circuit hypergraphs are sparse, and any move's gain is bounded between plus and minus the maximal vertex (G_{max})degree in the hypergraph (times the maximal hyperedge weight, if weights are used). This allows prioritizing moves by their gains. All affected gains can be updated in amortized-constant time, giving overall linear complexity per pass [17]. In [17] all moves with the same gain are stored in a linked list representing

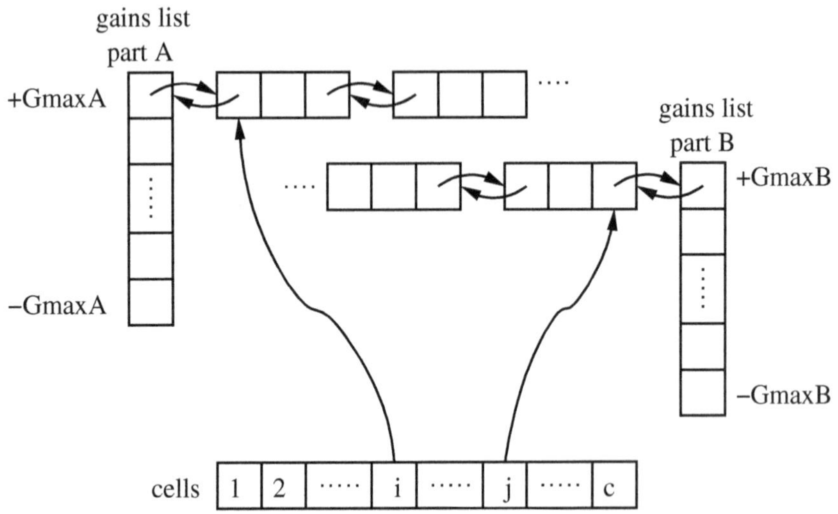

Fig. 4.5 The gain bucket structure as illustrated in [17]

a *gain bucket*. Figure 4.5 presents the gain bucket list structure. It is important to note that some gains G may be negative, and as such, FM performs hill-climbing and is not strictly greedy.

4.4.2.2 Multilevel Partitioning

The multilevel hypergraph partitioning framework was successfully verified in 1997 by [20, 21] and leads to the best known partitioning results ever since. The main advantage of Multilevel partitioning over flat partitioners is its ability to search the solution space more effectively by spending comparatively more effort on smaller coarsened hypergraphs. Good coarsening algorithms allow for high correlation between good partitioning for coarsened hypergraphs and good partitioning for the initial hypergraph. Therefore, a thorough search at the top of the multilevel hierarchy is worthwhile because it is relatively inexpensive when compared to flat partitioning of the original hypergraph, but can still preserve most of the possible improvement. The result is an algorithmic framework with both improved runtime and solution quality over a completely flat approach. Pseudo-code for an implementation of the multilevel partitioning framework is given in Algorithm 3.

Algorithm 3: Pseudo-code for the Multilevel Partitioning algorithm [19]

```
level = 0;
hierarchy[level] = hypergraph;
while hierarchy[level].vertex_count() less than 200 do
    next_level = cluster(hierarchy[level]);
    level = level + 1;
    hierarchy[level] = next_level;
end
partitioning[level] = a random initial solution for top-level hypergraph;
FM(hierarchy[level], partitioning[level]);
while level > 0 do
    level = level – 1;
    partitioning[level] = project(partitioning[level + 1], hierarchy[level]);
    FM(hierarchy[level], partitioning[level]);
end
```

As illustrated in Fig. 4.6, multilevel partitioning consists of 3 main components: clustering, top-level partitioning and refinement or *uncoarsening*. During clustering, hypergraph vertices are combined into clusters based on connectivity, leading to a smaller, clustered hypergraph. This step is repeated until obtaining only several hundred clusters and a hierarchy of clustered hypergraphs. We describe this hierarchy with the smaller hypergraphs being *higher* and the larger hypergraphs being *lower*. The smallest (top-level) hypergraph is partitioned with a very fast initial solution

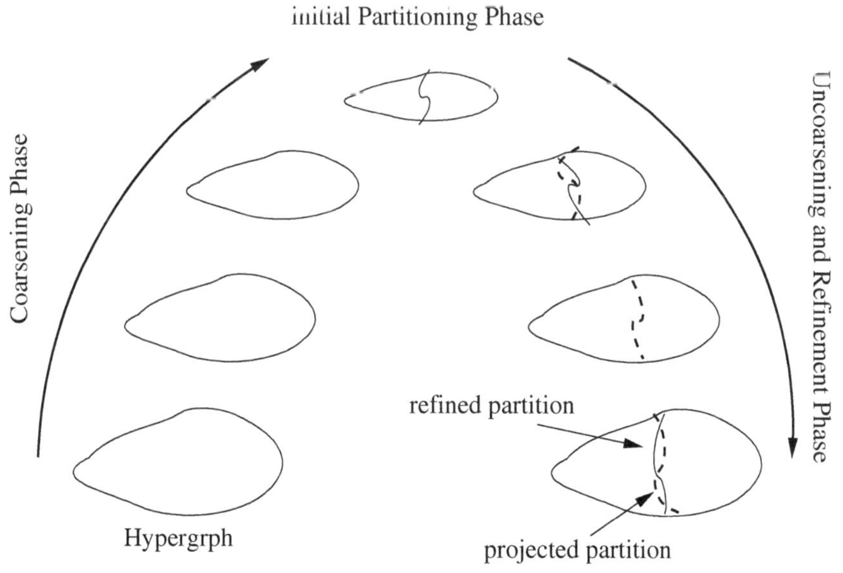

Fig. 4.6 Multilevel hypergraph bisection

generator and improved iteratively, for example, using the FM algorithm. The resulting partitioning is then interpreted as a solution for the next hypergraph in the hierarchy. During the refinement stage, solutions are projected from one level to the next and improved iteratively. Additionally, the *hMETIS* partitioning program [21] introduced several new heuristics that are incorporated into their multilevel partitioning implementation and are reportedly performance critical.

Algorithm 4: Generic simulated annealing-based placer [13]

```
S = RandomPlacement();
T = InitialTemperature();
R_limit = InitialR_limit;
while ExitCriterion() == false do
    while InnerLoopCriterion() == false do
        S_new = GenerateViaMove(S, R_limit);
        ΔC = Cost(S_new) − Cost(S);
        r = random(0,1);
        if r < e^(−ΔC/T) then
            S = S_new;
        end
    end
    T = UpdateTemp();
    R_limit = UpdateR_limit();
end
```

4.5 Placement

Placement algorithms determine which logic block within an FPGA should implement the corresponding logic block (instance) required by the circuit. The optimization goals consist in placing connected logic blocks close together to minimize the required wiring (wire length-driven placement), and sometimes to place blocks to balance the wiring density across the FPGA (routability-driven placement) or to maximize circuit speed (timing-driven placement). The 3 major classes of placers in use today are min-cut (Partitioning-based) [22, 23], analytic [24, 25] which are often followed by local iterative improvement, and simulated annealing based placers [26, 27]. To investigate architectures fairly we must make sure that our CAD tools are attempting to use every FPGA's feature. This means that the optimization approach and goals of the placer may change from architecture to architecture. Partitioning and simulated annealing approaches are the most common and used in FPGA CAD tools. Thus we focus on both techniques in the sequel.

4.5.1 Simulated Annealing Based Approach

Simulated annealing mimics the annealing process used to cool gradually molten metal to produce high-quality metal objects [26]. Pseudo-code for a generic simulated annealing-based placer is shown in Algorithm 4. A cost function is used to evaluate the quality of a given placement of logic blocks. For example, a common cost function in wirelength-driven placement is the sum over all nets of the half perimeter of their bounding boxes. An initial placement is created by assigning logic blocks randomly to the available locations in the FPGA. A large number of move, or local improvements are then made to gradually improve the placement. A logic block is selected at random, and a new location for it is also selected randomly. The change in cost function that results from moving the selected logic block to the proposed new location is computed. If the cost decreases, the move is always accepted and the block is moved. If the cost increases, there is still a chance to accept the move, even though it makes the placement worse. This probability of acceptance is given by $e^{-\frac{\Delta C}{T}}$, where ΔC is the change in cost function, and T is a parameter called temperature that controls probability of accepting moves that worsen the placement. Initially, T is high enough so almost all moves are accepted; it is gradually decreased as the placement improves, in such a way that eventually the probability of accepting a worsening move is very low. This ability to accept hill-climbing moves that make a placement worse allows simulated annealing to escape local minima of the cost function.

The R_{limit} parameter in Algorithm 4 controls how close are together blocks must be to be considered for swapping. Initially, R_{limit} is fairly large, and swaps of blocks far apart on a chip are more likely. Throughout the annealing process, R_{limit} is adjusted to try to keep the fraction of accepted moves at any temperature close to 0.44. If the fraction of moves accepted, α, is less than 0.44, R_{limit} is reduced, while if α is greater than 0.44 R_{limit} is increased. In [13], the objective cost function is a function of the total wirelength of the current placement. The wirelength is an estimate of the routing resources needed to completely route all nets in the netlist. Reductions in wirelength mean fewer routing wires and switches are required to route nets. This point is important because routing resources in an FPGA are limited. Fewer routing wires and switches typically are translated also into reductions of the delay incurred in routing nets between logic blocks. The total wirelength of a placement is estimated using a semi-perimeter metric, and is given by Eq. 4.2. N is the total number of nets in the netlist, $bbx(i)$ is the horizontal span of net i, $bby(i)$ is its vertical span, and $q(i)$ is a correction factor. Figure 4.7 illustrates the calculation of the horizontal and vertical spans of a hypothetical net that has 6 terminals.

$$WireCost = \sum_{i=1}^{N} q(i) \times (bb_x(i) + bb_y(i)) \tag{4.2}$$

The temperature decrease rate, the exit criterion for terminating the anneal, the number of moves attempted at each temperature (InnerLoopCriterion), and the method

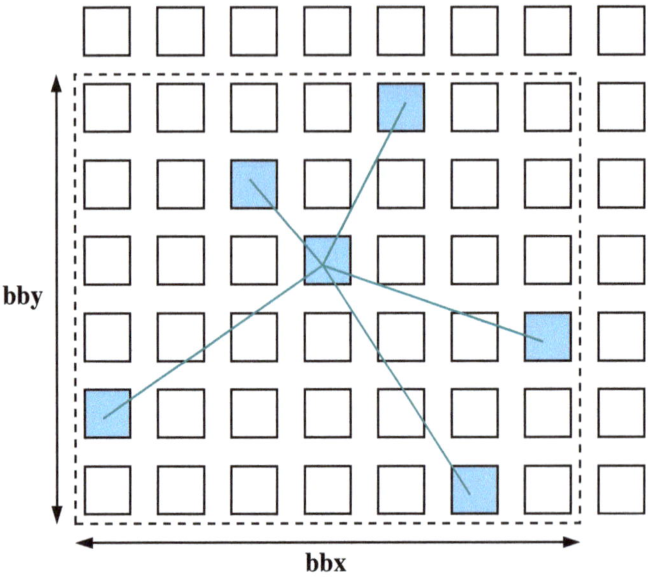

Fig. 4.7 Bounding box of a hypothetical 6-terminals net [13]

by which potential moves are generated are defined by the annealing schedule. An efficient annealing schedule is crucial to obtain good results in a reasonable amount of CPU time. Many proposed annealing schedules are *fixed* schedules with no ability to adapt to different problems. Such schedules can work well within the norrow application range for which they are developed, but their lack of adaptability means they are not very general. In [28] authors propose an *adaptive* annealing schedule based on statistics computed during the anneal itself. Adaptive schedules are widely used to solve large scale optimization problems with many variables.

4.5.2 Partitioning Based Approach

Partitioning-based placement methods, also referred to as min-cut methods, are based on graph partitioning algorithms such as the Fiduccia-Mattheyses (FM) Algorithm [17], and Kernighan Lin (KL) algorithm [22]. Partitioning-based placement are suitable to tree-based FPGA architectures. The partitioner is applied recursively to each hierarchical level to distribute netlist cells between clusters. The aim is to reduce external communications and to collect highly connected cells into the same cluster.The partitioning-based placement is also used in the case of Mesh-based FPGA. The device is divided into two parts, and a circuit partitioning algorithm is applied to determine the adequate part where a given logic block must be placed to minimize the number of cuts in the nets that connect the blocks between partitions, while leaving highly-connected blocks in one partition. A divide-and-conquer strategy is used in these heuristics. By partitioning the problem into sub-parts, a drastic reduction in search space can be achieved. On the whole, these algorithms perform in the top-down manner, placing blocks in the general regions which they should belong to. In the Mesh FPGA case, partitioning-based placement algorithms are good from a *global* perspective, but they do not actually attempt to minimize wirelength. Therefore, the solutions obtained are sub-optimal in terms of wirelength. However, these classes of algorithms run very fast. They are normally used in conjunction with other search techniques for further quality improvement. Some algorithms [29] and [30] combine multi-level clustering and hierarchical simulated annealing to obtain ultra-fast placement with good quality. In our project, the partitioning-based placement approach is used for the placement of Tree-based FPGA architectures.

4.6 Routing

The FPGA routing problem consists in assigning nets to routing resources such that no routing resource is shared by more than one net. Pathfinder [31] is the current, state-of-the-art FPGA routing algorithm. Pathfinder operates on a directed graph abstraction ($G(V, E)$) of the routing resources in an FPGA. The set of vertices V in the graph represents the IO terminals of logic units and the routing wires in the interconnect structure. An edge between two vertices represents a potential connection between these two vertices. Figure 4.8 presents a part of a routing graph in a Mesh-based interconnect. Given this graph abstraction, the routing problem for a given net is to find a directed tree embedded in G that connects the source terminal of the net to each of its sink terminals. Since the number of routing resources in an FPGA is limited, the goal of finding unique, non-intersecting trees for all the nets in a netlist is a difficult problem. $Pathfinder$ uses an iterative, negotiation-based approach to successfully route all the nets in a netlist. During the first routing iteration, nets are freely routed

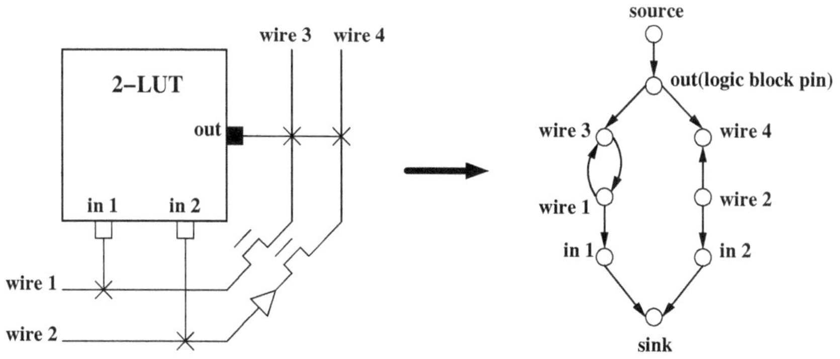

Fig. 4.8 Modelling FPGA routing architecture as a directed graph [13]

without paying attention to resource sharing. Individual nets are routed using Dijkstra s shortest path algorithm [32]. At the end of the first iteration, resources may be congested because multiple nets have used them. During subsequent iterations, the cost of using a resource is increased, based on the number of nets that share the resource, and the history of congestion on that resource. Thus, nets are made to negotiate for routing resources. If a resource is highly congested, nets which can use lower congestion alternatives are forced to do so. On the other hand, if the alternatives are more congested than the resource, then a net may still use that resource. The cost of using a routing resource n during a routing iteration is given by Eq. 4.3.

$$c_n = (b_n + h_n) \times p_n \tag{4.3}$$

where b_n is the base cost of using the resource n, h_n is related to the history of congestion during previous iterations, and p_n is proportional to the number of nets sharing the resource in the current iteration. The p_n term represents the cost of using a shared resource n, and the h_n term represents the cost of using a resource that has been shared during earlier routing iterations. The latter term is based on the intuition that a historically congested node should appear expensive, even if it is slightly shared currently. The Pseudo-code of the *Pathfinder* routing algorithm is presented in Algorithm 5.

Algorithm 5: Pseudo-code of the Pathfinder routing algorithm [31]

```
Let: RT_i be the set of nodes in the current routing of net i
while shared resources exist do
    /*Illegal routing*/
    foreach net, i do
        rip-up routing tree RT_i;
        RT(i) = s_i foreach sink t_ij do
            Initialize priority queue PQ to RT_i at cost 0;
            while sink t_ij not found do
                Remove lowest cost node m from PQ;
                foreach fanout node n of node m do
                    Add n to PQ at PathCost(n) = c_n + PathCost(m);
                end
            end
            foreach node n in path t_ij to s_i do
                /*backtrace*/
                Update c_n;
                Add n to RT_i;
            end
        end
    end
    update h_n for all n;
end
```

An important measure of routing quality produced by an FPGA routing algorithm is the critical path delay. The critical path delay of a routed netlist is the maximum delay of any combinational path in the netlist. The maximum frequency at which a netlist can be clocked has an inverse relationship with critical path delay. Thus, larger critical path delays slow down the operation of netlist. Delay information is incorporated into Pathfinder by redefining the cost of using a resource n (Eq. 4.4).

$$c_n = A_{ij} \times d_n + (1 - A_{ij}) \times (b_n + h_n) \times p_n \tag{4.4}$$

The c_n term is from Eq. 4.3, d_n is the delay incurred in using the resource, and A_{ij} is the criticality given by Eq. 4.5.

$$A_{ij} = \frac{D_{ij}}{D_{max}} \tag{4.5}$$

D_{ij} is the maximum delay of any combinational path going through the source and sink terminals of the net being routed, and D_{max} is the critical path delay of the netlist. Equation 4.4 is formulated as a sum of two cost terms. The first term in the equation represents the delay cost of using resource n, while the second term represents the congestion cost. When a net is routed, the value of A_{ij} determines whether the delay or the congestion cost of a resource dominates. If a net is near critical (i.e. its A_{ij} is close to 1), then congestion is largely ignored and the cost of using a resource is primarily determined by the delay term. If the criticality of a net is low, the congestion term in Eq. 4.4 dominates, and the route found for the net avoids congestion while potentially incurring delay. Pathfinder has proved to be one of the most powerful FPGA routing algorithms to date. Pathfinder's negotiation-based framework that trades off delay for congestion is an extremely effective technique for routing signals on FPGAs. More importantly, Pathfinder is a truly architecture-adaptive routing algorithm. The algorithm operates on a directed graph abstraction of an FPGA's routing structure, and can thus be used to route netlists on any FPGA that can be represented as a directed routing graph.

4.7 Two-Dimensional CAD for Tree-Based Architecture

In this thesis we are exploring different FPGA architecture topologies, we developed CAD tools as generic as possible to deal with different types of FPGA architectures. We propose a set of generic tools requiring a minimum effort to be adapted to a specific architecture topology. In Fig. 4.9 we present the dependency between each phase in the CAD flow and the target architecture.

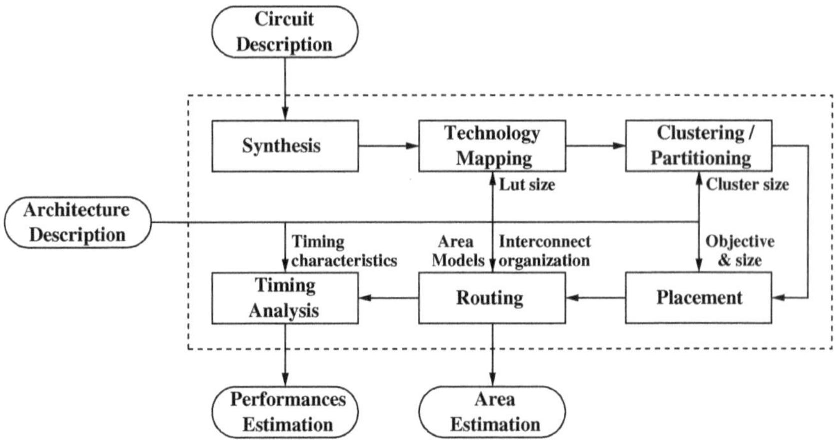

Fig. 4.9 Architectures exploration platform

4.7.1 Synthesis and Mapping

Synthesis consists in translating a circuit description into a gate-level representation. As illustrated on Fig. 4.9 this operation is architecture independent. In our flow we use SIS [1] synthesis tool. It can be replaced by any other commercial synthesis tool. As explained in Sect. 4.2, mapping consists in translating the description based on a boolean logic gates into a description with k-input LUTs and flip-flops. The only required architecture parameter is k, the LUT inputs number. In our flow we use FlowMap algorithm [2], which is included in SIS package. As presented in Fig. 4.9, this tool depends only on LUTs size and can target any interconnect topology. It can be driven by different objectives like timing (depth optimization) and area (LUTs number). Notice that today FPGA commercial mapping tools can target specific architectures interconnects. Thus in this early stage they can alleviate routing congestion and improve performance.

4.7.2 Clustering and Partitioning

In general, FPGA architectures, the programmable interconnect is organized in multiple hierarchical levels. Hierarchy becomes an interesting feature to improve density, to reduce run time effort (divide and conquer) and to consider local communication. For example in the case of Mesh-based industrial FPGA architecture, interconnect is organized in two-levels of hierarchy: (1) Mesh level where clusters are surrounded by depopulated interconnect organized in row and columns and (2) cluster level where LBs are connected using a full cross bar. Stratix architecture has also the same number of hierarchical levels but with a more optimized interconnect topology [33]. In the case of a Tree-based interconnect we get multiple hierarchical levels. Level numbers depends on the total number of LBs and clusters size (arity). Basically when two signals are within the same hierarchy level, it does not really matter where within that hierarchy they are. Similarly, geometrically close cells incur greater delay to get to other locations outside their hierarchical boundary than to distant cells within their hierarchical boundary. Thus, unlike flat or island style Mesh-based FPGA, a hierarchical architecture uses a natural placement algorithm based on recursive partitioning. The netlist instance are partitioned between architecture clusters in the best possible way using the newly developed CAD Flow with the careful consideration of multilevel hierarchical organization of Tree-based interconnect to reduce the desired partitioning objectives. For this purpose we implemented 3 different partitioning objectives:

1. CUT:- Corresponds to the total number of nets crossing parts boundaries.
2. SOED:- Sum Of External Degree: External part degree corresponds to the number nets crossing the part boundary.
3. MED:- Maximum External Degree: Corresponds to the maximum degree over all parts.

These objectives can be combined or considered separately. The Fig. 4.10 illustrate an example of partitioning and evaluation of the 3 different partitioning objectives. For Tree-based FPGA architecture we developed two main partitioning approaches: bottom up (clustering) and top-down. The choice between both approaches depends on levels number, clusters size, clusters number in each level and problem constraints. For example *t-vpack* [10] a bottom-up clustering tool is used to construct clusters in the case of Mesh-based FPGA architecture. We replaced *t-vpack* with *hMetis* [20, 34] a top-down partitioner. By using top-down partitioner [35], we observed improvements in the reduction of external nets at the cost of increasing run time. In fact top-down approaches based on FM refinement heuristics are efficient when we target a small number of clusters (parts) with an important size (balance constraint) [17]. Conversely in the case of Mesh-based architecture clusters size is small (between 4 and 16) and clusters number is in general important. To investigate

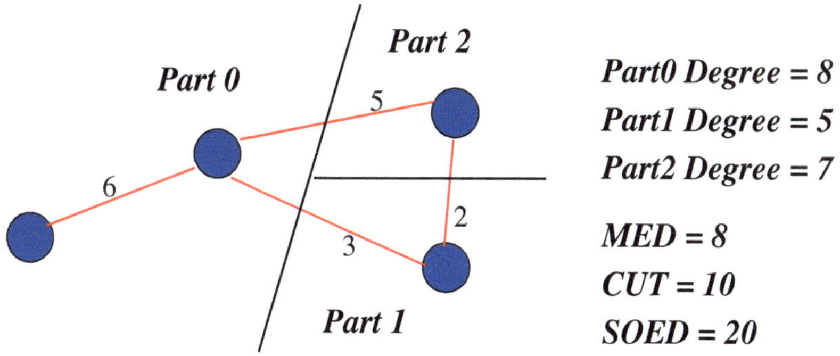

Fig. 4.10 Illustration and evaluation of different partitioning objectives

partitioning approaches we used a Multilevel hypergraph structure called *Mangrove*. *Mangrove* provides a development framework for efficient modeling of hypergraph nested partitions. It offers a compact C++ data structure and high level API and this structure is organized as follows [35]:

- *ClusteringHierarchy*: holds a vector of nested partitions called *Clustering Level*, and refers to a unique enclosing cluster *TopLevelCluster*,
- *ClusteringLevel*: corresponds to the set of clusters at the partitioning at a given level. A *clusteringLevel* corresponds to a Hypergraph where nodes are clusters situated at this level.
- *Cluster*: Aggregates sub-clusters belonging to a lower *ClusteringLevel* (unless leaf one). A *Cluster* may cross multiple levels and has *UpperLevel* and *Lower Level* identifiers,
- *Net*: presents a tree of branches,
- *Branch*: represents the net (signal) crossing point of a cluster boundary. Branch bifurcates within a cluster if the net crosses at least 2 sub-clusters.

Since in Mangrove a *clusteringLevel* can be added at any level, this structure can be used in different partitioning approaches: Bottom-up and top-down. The combination of both approaches leads to an efficient multilevel partitioner where first multilevel bottom-up coarsening is run and then top-down multilevel refinement is applied. In Fig. 4.11 we show the different steps of recursive netlist partitioning based on a multilevel approach. The netlist is first partitioned into 2 parts (first level) and then instances inside each part are partitioned into 2 fractions. In each partitioning phase

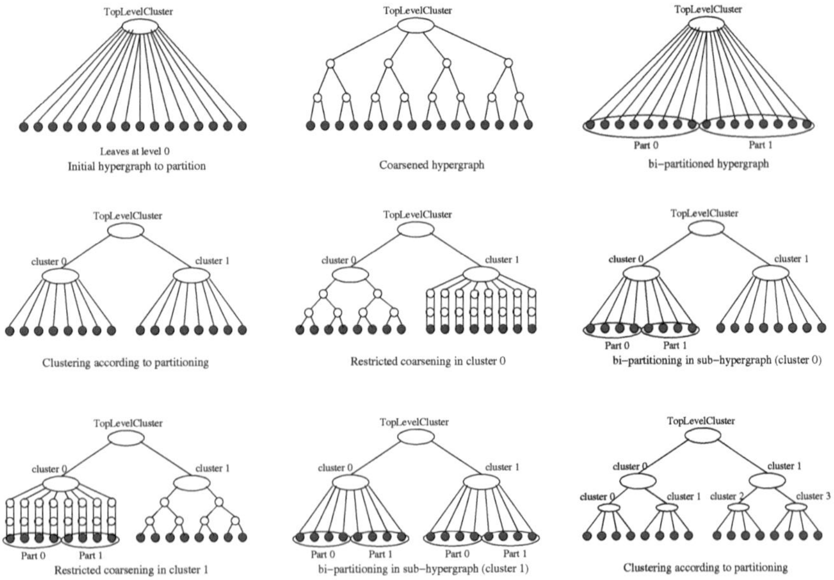

Fig. 4.11 2 levels recursive bi-partitioning steps

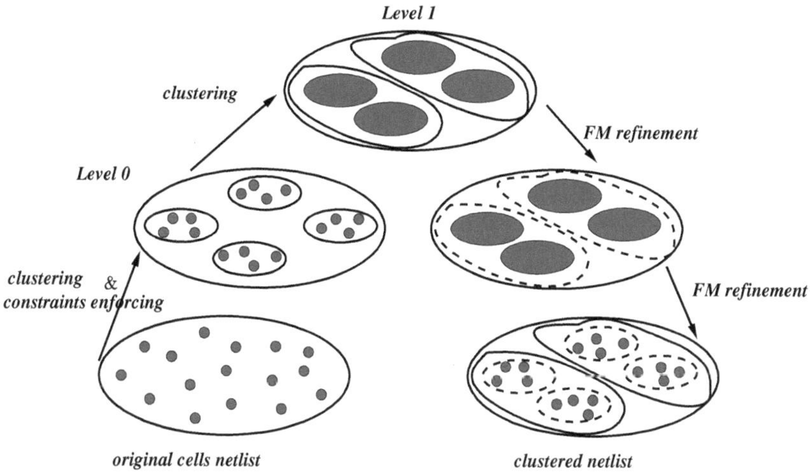

Fig. 4.12 Multilevel clustering and refinement

we apply a multilevel coarsening followed by a multilevel refinement as illustrated in Fig. 4.12. After completing the clustering phase (with Pins-limit strategy), we obtain a tree of clusters each one containing k sub-clusters. During the refinement phase, cells will be moved between clusters (parts) to optimize an objective function without violating the constraints imposed by the cluster size. In a level ℓ, cells are not allowed to move between all clusters, because this can decrease the quality of the solution obtained in the higher level. To prevent such unwanted effect, cells can only move between neighboring clusters. We call neighboring clusters, all clusters in a level belonging to the same supercluster. Thus in every level, neighboring clusters will be isolated and form a subgraph. In Fig. 4.12 those subgraphs are represented by the continuous lines and partition by the dashed ones. A cell is allowed only to move across dashed lines. The objective function is specific to each subgraph and corresponds to the Maximum External Degree (MED) of all parts belonging to the same subgraph. An FM algorithm [17] is applied to a subgraph to optimize the local objective function. The complexity of our k-way refinement is reduced since we apply it successively for each subgraph (in each subgraph there are small number of parts: Arity of the architecture) and only for the highest levels (where pins-limit strategy fails). Finally, we obtain the partitioning result corresponding to each level. The final result describes how instances are distributed between clusters of the Tree-based topology. Recursive partitioning is also interesting to reduce run time since it allows to avoid applying FM heuristics directly on a large number of parts, which can dramatically increase the partitioning run time.

4.8 Timing Analysis

Timing analysis evaluates performances of a circuit implemented on a FPGA in terms of functional speed. Thus, once an application is completely placed and routed we estimate the minimum feasible clock to run it. To achieve timing analysis we need 2 different graphs:

- Routed graph: Describes the way netlist instances are routed using architecture resources. This graph allows to evaluate routing delays between netlist instances connections. A path connecting two instances crosses several wires and switches. The connection delay is equal to the sum of resources delays.
- Timing graph: It is a direct acyclic graph generated from the netlist hypergraph. Nodes correspond to instances pins and edges to connections. Based on the resulting routed graph, each edge is labeled with the corresponding routed connection delay. The minimum required clock period is determined via a breadth-first traversal applied on this graph.

Only the routed graph is architecture dependent. Timing graph generation and critical path extraction depend only on netlist to implement. First the circuit under consideration is presented as a directed graph. Nodes in the graph represent input and output pins of circuit elements such as LUTs, registers, and I/O pads. Connections between these nodes are modeled with edges in the graph. Edges are added between the inputs of combinational logic Blocks (LUTs) and their outputs. These edges are annotated with a delay corresponding to the physical delay between the nodes. Register input pins are not joined to register output pins. To determine the delay of the circuit, a breadth first traversal is performed on the graph starting at sources (input pads, and register outputs). Then the arrival time, $T_{arrival}$, at all nodes in the circuit is computed with the following equation

$$T_{arrival}(i) = \max_{j \in fanin(i)} \{T_{arrival}(j) + delay(j, i)\}$$

where node i is the node currently being computed, and $delay(j, i)$ is the delay value of the edge joining node j to node i. The delay of the circuit is then the maximum arrival time, D_{max}, of all nodes in the circuit. To guide a placement or routing algorithm, it is useful to know how much delay may be added to a connection before the path that the connection is on becomes critical. The amount of delay that may be added to a connection before it becomes critical is called the slack of that connection. To compute the slack of a connection, one must compute the required arrival time, $T_{required}$, at every node in the circuit. We first set the $T_{required}$ at all sinks (output pads and register inputs) to be D_{max}. Required arrival time is then propagated backwards starting from the sinks with the following equation:

$$T_{required}(i) = \min_{j \in fanout(i)} \{T_{required}(j) - delay(j, i)\}$$

Finally, the slack of a connection (i,j) driving node, j, is defined as:

$$Slack(i, j) = T_{required}(j) - T_{arrival}(i) - delay(i, j)$$

4.9 Summary

The most important architectural feature of an FPGA is the interconnect structure. During architectures exploration, the effectiveness of an FPGA interconnect structure is evaluated using placement and routing tools. Fortunately, some classes of the used algorithms are architecture-adaptive and can be used to evaluate different structures. In the next chapter we will present an exploration tools platform that can be adapted to test different target architecture topologies for the implementation 3D Tree-based FPGA.

References

1. E.M. Sentovich, K.J. Singh, L. Lavango, C. Moon, R. Murgai, A. Saldanha, H. Savoj, P. Stephan, R.K. Brayton, A. Sangiovanni-Vincentelli, SIS: A System for Sequential Circuit Synthesis, Technical Report No. UCB/ERL, M92/41. (University of California Berkeley, 1992)
2. J. Cong, Y. Ding, FlowMap: An optimal technology mapping algorithm for delay optimization in look-up-table based fpga designs. IEEE Trans. Comput. Aided Des, 1–12 (1994)
3. J. Cong, Y. Ding, On area/depth trade-off in LUT-based FPGA technology mapping. IEEE Trans. VLSI Syst. **2**(2), 137–148 (1994)
4. J. Cong, Y. Hwang, Simultaneous depth and area minimization in LUT-based FPGA mapping, in *ACM/SIGDA International Symposim on FPGAs*, pp. 68–74 (1995)
5. J. Cong, Y. Ding, Structural gate decomposition for depth-optimal technology in LUT-based FPGA designs. ACM Trans. Des. Autom. Electr. Syst. **5**(3) (2000)
6. D. Huang, A. Kahng, When clusters meet partitions: new density based methods for circuit decomposition, in *IEEE European Design and Test Conference*, pp. 60–64 (1995)
7. L. Hagen, A. Kahng, Combining problem reduction and adaptive multi-start: a new technique for superior iterative partitioning, in *IEEE Transactions on Computer-Aided Design*, pp. 92–98 (1997)
8. R. Murgai, R. Brayton, A. Sangiovanni-Vincentelli, On clustering for minimum delay/area, in *IEEE International Conference on Computer Aided Design*, pp. 6–9 (1991)
9. M. Dehkordi, S. Brown, The effect of cluster packing and node duplication control in delay driven clustering, in *IEEE International Conference on Field Programmable Technology*, pp. 227–233 (2002)
10. A. Marquart, V. Betz, J. Rose, Using cluster-based logic block and timing-driven packing to improve fpga speed and density, in *ACM International Symposium on FPGA, Monterey*, pp. 37–46 (1999)
11. A. Singh, M. Marek-Sadowska, Efficient circuit clustering for area and power reduction in FPGAs, in *International Symposium on Field Programmable Gate Arrays*, pp. 59–66 (2002)
12. E. Bozorgzadeh et al., Routability-driven packing: Metrics and algorithms for cluster-based FPGAs. IEEE J. Circuits Syst. Comput. **13**(1), 77–100 (2004)
13. V. Betz, J. Rose, A, *Marquardt Architecture and CAD for Deep Sub-micron FPGAs*. (Kluwer, Norwell, MA, 1999)
14. M.R. Garey, D.S. Johnson, L. Stockmeyer, Some simplified NP complete problems, in *Sixth Annual ACM Symposium on Theory of Computing*, pp. 47–63 (1974)
15. M.R. Garey, D.S. Johnson, *Computers and Intractability: A Guide to the Theory of NP-Completeness, an Francisco* (Freeman, CA, 1979)
16. B. Kernighan, S. Lin, An efficient heuristic procedure for partitioning graph. Bell Syst. Technol. J. **49**, 299–307 (1970)

17. C.M. Fiduccia, R.M. Mattheyeses, A liner-time heuristic for improving network partitions, in *Design Automation Conference*, vol. 7, pp. 175–181 (1982)
18. T. Bui, S. Chaudhuri, T. Leighton, M. Sipser, Graph bisection algorithms with good average behavior. Combinatorica (1987)
19. D.A. Papa, I.L. Markov, *Hypergraph Partitioning and Clustering, Technical Report*. (University of Michigan, EECS Department, 2007)
20. G. Karypis, R. Aggarwal, V. Kumar, S. Shekhar, Multilevel hypergraph partitioning: application in VLSI design, in *ACM, Design Automation Conference*, pp. 526–529 (1997)
21. G. Karypis, V. Kumar, Multilevel k-way hypergraph partitioning, in *Proceedings of the 36th annual ACM/IEEE Design Automation Conference*, pp. 343–348 (1999)
22. A. Dunlop, B. Kernighan, A procedure for placement of standard-cell VLSI circuits. IEEE Trans. CAD, 92–98 (1985)
23. D. Huang, A. Kahng, Partitioning-based standard-cell global placement with an exact objective, in *ACM Symposium on Physical Design*, pp. 18–25 (1997)
24. G. Sigl, K. Doll, F. Johannes, Analytical placement: a linear or a quadratic objective function?, in *ACM Design Automation Conference*, pp. 427–432 (1991)
25. C. Alpert, T. Chan D. Huang A. Kahng I. Markov P. Mulet, K. Yan, Faster Minimization of linear wire-length for global placement, in *ACM Symposium on Physical Design*, pp. 4–11 (1997)
26. S. Kirkpatrick, C.D. Gelatt, M.P. Vecchi, Optimization by simulated annealing. Science **220**, 671–680 (1983)
27. C. Sechen, A. Sangiovanni-Vincentelli, *The Timberwolf Placement and Routing Package*. (JSSC, 1985) pp. 510–522
28. M. Huang, F. Romeo, A. Sangiovanni-Vincentelli, An efficient general cooling schedule for simulated annealing, in *Proceedings of ICCAD*, pp. 381–384 (1986)
29. Y. Sanker, J. Rose, *Trading Quality for Compile Time: Ultra-Fast Placement for FPGAs*. ACM International Symposium on FPGAs, FPGA (1999)
30. P. Du, G. W. Grewal, S. Areibi, D.K. Banerji, A fast hierarchical approach to FPGA placement, in *ESA/VLSI*, pp. 497–503 (2004)
31. L. McMurchie, C. Ebeling, PathFinder: a negotiation based performance driven router for FPGAs, in *Conference of Field Programmable Gate Arrays FPGA*, vol. 12, no 2, pp. 291–301 (1995)
32. T. Cormen, C. Leiserson, R. Rivest, *Introduction to Algorithms*. (MIT Press, Cambridge, 1990)
33. D. Lewis et al., The stratix logic and routing architecture, in *International Symposium on Field Programmable Gate Arrays, FPGA-2003*, pp. 12–20, Feb 2003
34. N. Selvakkumaran, G. Karypis, Multi-objective hypergraph-partitioning algorithm for cut and maximum subdomain-degree minimization. IEEE Trans. Comput. Aided Des. Integr. Circuits **25**(3), 504–517 (2006)
35. Z. Marrakchi, H. Mrabet, H. Mehrez, Hierarchical FPGA clustering to improve routability, in *Ph.D. Research Conference in Microelectronics, PRIME* (2005)

Chapter 5
Three-Dimensional FPGAs: Configuration and CAD Development

Abstract The primary focus of this chapter is to demonstrate a 3D integration scheme to partition and optimize the multilevel programmable interconnect network of Tree-based FPGA based on Butterfly-Fat-Tree network topology, where TSVs are incorporated in active layers of the 3D chip. This chapter describes the details of the architecture of 3D FPGAs and state-of-the-art 3D technology for Mesh-based FPGAs. To take advantage of 3D integrated circuits, it should be investigated how FPGA should be physically partitioned into different active layers. Proper physical partitioning has a great impact on the performance improvement of the system. This chapter discuss different partitioning schemes and design techniques and associated 3D CAD tools of 3D FPGAs.

5.1 Introduction

Modern Field Programmable Gate Arrays (FPGAs) have become a viable alternative to cell based design technology by providing re-configurable computing platforms with improved performance and density. With the onset of sub-100 nm CMOS technologies, the design and prototyping cost of the cell-based custom integrated circuit implementation have become exorbitant for most ASICs, making FPGAs increasingly popular. With their regular structure, they also scale easily with sub-100 nm technologies. Current FPGAs, however cannot meet the area and speed requirements of many ASICs due to their high programming overhead. Recent studies shows that in Mesh-based industrial FPGAs, 80 % of overall design delay and 90 % of the chip area are attributed to programmable routing resources [1–3]. It has also reported that in Mesh-based FPGA, programmable interconnects contributes as much as 60 % of the total dynamic power consumption [4]. Considering the area, delay and power consumption, the programmable interconnects are the key elements in FPGA design [3, 5].

V. Pangracious et al., *Three-Dimensional Design Methodologies for Tree-based FPGA Architecture*, Lecture Notes in Electrical Engineering 350, DOI 10.1007/978-3-319-19174-4_5

The 3D integrated circuit (IC) technology has emerged as one of the most promising solutions for overcoming the challenges in interconnection and integration complexity in modern circuit designs [6]. The 3D integration technology can effectively reduce global interconnect length and increase circuit performance without increasing the power consumption. Through Silicon Vias (TSVs) are the key enabling technology element for 3D integration, which is currently being actively evaluated as a potential solution to reduce the interconnect delay and increase the logic density in FPGA. Recently, multiple technology and product demonstrations of TSV in silicon interposer have been reported for high performance FPGA applications [7–9]. Based on the design and manufacturing specifications and maturity of enabling technologies, a pattern in technology adoption is beginning to emerge. Three-dimensional integration where TSVs are incorporated in active device layers, is considered to be the Holy-Grail of vertical die stacking. However the recent product demonstrations from major FPGA manufactures revels the adoption of silicon interposer based 3D integration scheme, where TSVs are incorporated in passive silicon interposer [8, 9]. The primary focus of this chapter is to demonstrate a 3D integration scheme to partition and optimize the multilevel programmable interconnect network of Tree-based FPGA based on BFT network topology, where TSVs are incorporated in active layers of the 3D chip. This chapter describes the details of the architecture of 3D FPGAs and state-of-the-art 3D technology for Mesh-based FPGAs.

5.2 3D FPGA Architectures: An Overview

To take advantage of 3D integrated circuits, it should be investigated how FPGA should be physically partitioned into different active layers. Proper physical partitioning has a great impact on the performance improvement of the system. As in any FPGA, the 3D FPGA consists of programmable LBs that are connected by an interconnect, which are both programmed by the configuration memory. The 3D interconnect methodology is quite different compared to the 2D interconnect system. Figure 5.1 illustrates the possible architecture for 3D Mesh-based FPGA. In this architecture different active layers have the same structure, all layers have configurable logic blocks (LBs) and switch blocks (SBs). The connection between blocks is similar in all active layers. The advantage is that LBs that were far away in a 2D FPGA are now much closer at the upper or lower layers and provides more degrees of freedom during the routing since the third dimension can be employed [10].

5.2.1 FPGA Die Stacking

In FPGAs, the configuration memory blocks provides predefined constant voltages to the functional circuit. These constant voltages or memory values are not subject of any timing requirement and can be placed in separate layer with no overhead.

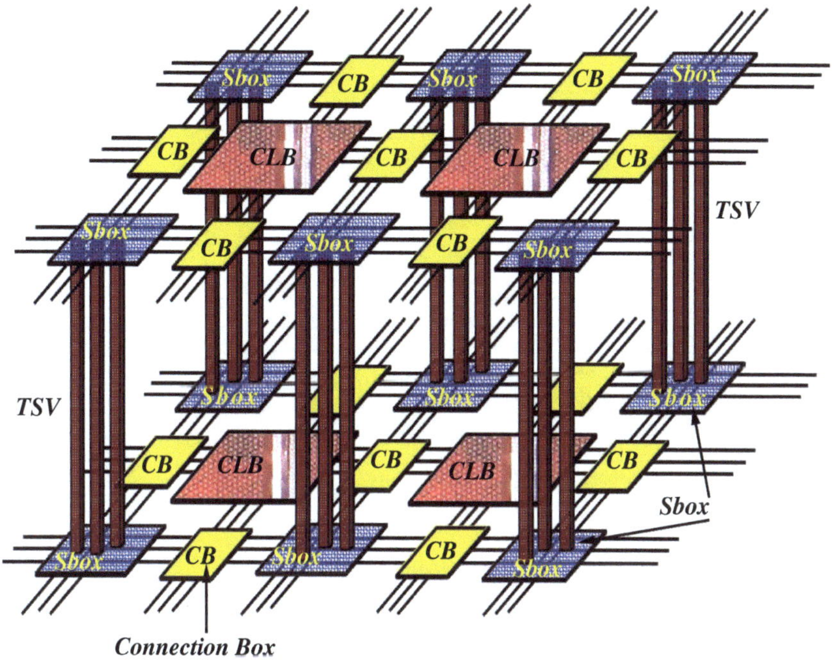

Fig. 5.1 A typical mesh-based 3D FPGA architecture with 3D SBs using vertical interconnections (TSVs)

The second possibility of implementing 3D FPGA architecture is based on placing the FPGA configuration memory on a different layer as illustrated in Fig. 5.2. In this architecture, one active layer is specially devoted to configuration memory cells and another active layers is devoted to programmable LBs and SBs. This approach provides more flexibility to increase logic density of FPGA as well as to reduce the area of the tile by a factor of two and the length of interconnect would reduce by square root of two. The delay and power dissipation of interconnect would also reduce because of the shorter capacitance on interconnect. However, current technology requires a large inter-tier via (TSV) for 3D interconnection to other tiers. The size of the via depends on the technology and accuracy of aligning two dies during the bonding process. A tile may have few thousands configuration cells and if all configuration cells are placed in another tier, the area overhead of 3D connections will be too high. Depending on the technology, the area overhead of 3D interconnect may even exceed the original area of the die and the size of an inter-tier via scales with the accuracy of the die alignment.

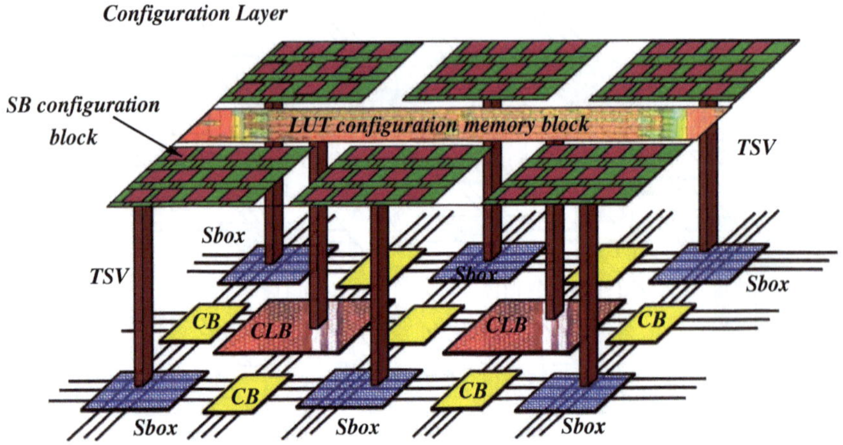

Fig. 5.2 A typical mesh-based 3D FPGA architecture

5.2.2 Monolithic FPGA Implementation

A more promising 3D-IC approach for implementing such a 3D FPGA is monolithic stacking, whereby the active devices are lithographically built in between metal layers [11, 12]. The main advantage of such approach is that, in principle, it can achieve comparable vertical via density and scale at the same rate as the base CMOS technology. Although this approach is yet to be developed for FPGA application, there is much evidence that forming transistors on a dielectric with low thermal budget is quit feasible [13–15]. The process technology for the added layers can be much simpler than a full CMOS process. Specifically, the switch layer only needs one type of MOS transistors, while the memory layer can be implemented using two-transistor flash technology [16] or a programmable solid-electrolyte switch [17], both of which promise to achieve higher densities than static random-access memory (SRAM) with much simpler processes. The monolithic 3D FPGA process enables much higher vertical interconnect density than 3D stacked FPGAs. This requires a radically new 3D architecture than simple stacking of 2D FPGAs. For example, the simple 3D stacking methodology can be used to build high density FPGAs with fully 3D switch boxes [18], which were shown to achieve over 50 % reduction in channel width, interconnect delay, and power consumption over a baseline 2D FPGA. The 3D monolithic FPGA architecture however requires a significantly more complex 3D technology than 3D stacked FPGAs.

5.3 State-of-the-Art: 3D FPGA Implementation

To provide the required reconfigurable functionality, FPGAs provide a large amount of programmable interconnect resources in the form of wire segments, switches, and signal repeaters. Typically, the delay of FPGAs is dominated by these interconnects. Reducing the lengths of interconnects can lead to significant improvements in the performance of FPGAs. Moreover, these programmable interconnect resources typically consume a large portion of the FPGA silicon die area. Since die area is one of the main factors that determine manufacturing costs, reducing the silicon footprint can lead to significant improvements in the manufacturing costs of FPGAs. The advantages of 3D FPGAs have evoked significant interest, and several studies have looked at them in the past. A 3D FPGA that used package-level integration to stack multiple 2D Mesh-based FPGAs interconnected using solder bumps is presented in [19]. The minimum pitch of these vertical interconnection is $\approx 100\,\mu m$. An opto-electronic 3D FPGA is proposed in [20], in which the inter-tier communication is realized using optical links. The main advantage of using optical links is to improve performance by provide a large vertical channel density.

The Rothko 3D FPGA [21] is a 3D extension of the Triptysch sea-of-gate architecture [22]. Rothko is 3D design platform allows designers to stack 2D CMOS VLSI circuits to build 3D VLSI structures. This design platform provides flexibility is placing TSVs anywhere in the chip. For Rothko 3D FPGA, the 3D integration is done at wafer level and inter-tier communication is established using TSVs (metal vias). To stack tow active layers of the FPGA chip, Rothko used CMOS based bulk technology and Silicon-on-Insulator (SOI) technology. For a two-tier circuit, the first tier is processed using CMOS bulk technology and the second tier is processed using SOI technology. The problem associated with such a process is, it needs two different wafers and technology, which lead to increase in the cost of manufacturing. A dynamically reconfigurable 3D Mesh-based FPGA is presented in [23], which consisted of three physical layers: local routing and logic block layer, main routing layer, and memory layer. Such designs improve the flexibility and performance of FPGA at the cost of a large number of TSVs.

A conceptually appealing approach to closing the performance gap between FPGAs and custom cell-based ASICs is to stack the programming overhead of an FPGA on top of the LBs and implement the interconnect layers using monolithic state-of-the-art CMOS technology. The implementation and performance analysis of a monolithically integrated 3D FPGA presented in [12, 24]. Monolithic 3D integration provides very fine and high density vertical vias ($2B/cm^2$), which allows the integration technology to stack FPGA in three tiers, in which all SRAM cells were moved to a separate layer and monolithical stacking is used so that there is no overhead on inter-tier connections as illustrated in Fig. 5.3. In monolithical stacking, electronic components and their connections (wiring) are lithographically built in layers on a single wafer, hence no TSV is needed. In this case, the area overhead on TSVs is eliminated and the logic density is greatly improved due to stacking. Researches are also looked at theoretical models for 3D FPGAs. An analytical model for predicting

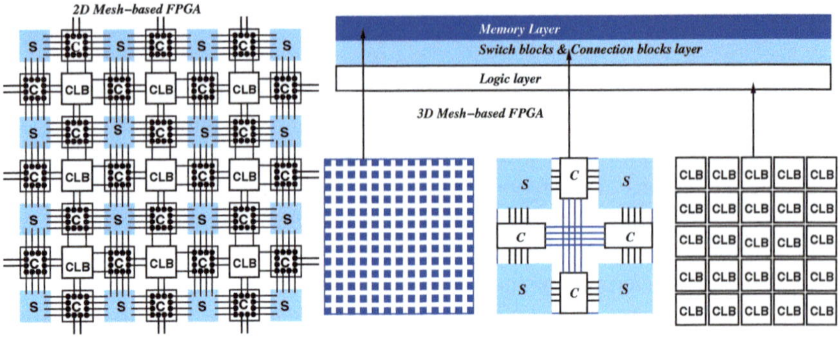

Fig. 5.3 A typical mesh-based 3D monolithic FPGA architecture

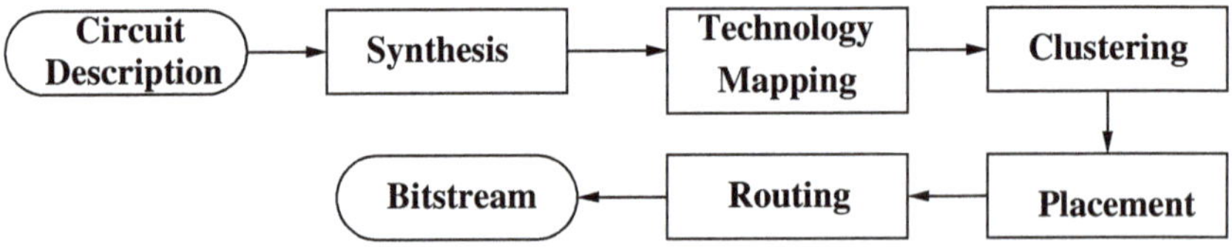

Fig. 5.4 Typical FPGA CAD flow

interconnect requirements in 3D FPGA presented in [18] to examined opportunities for 3D implementation of FPGAs. In their exploration, 3D (6-way) switching blocks were used and all the components in the FPGA were evenly distributed between layers in fine granularity. Experimental results on 0:25 μm technology showed that in FPGAs with 70 K logic cells, the improvement in LUT density can be 25–60 % in 3D implementation with two to four tiers. The interconnect delay is significantly reduced (50 %), and the clock frequency as 2D FPGA, the reduction in power dissipation is 35–55 %, when compared to 2D Mesh-based FPGAs. Recently this model extended to incorporate clustered logic blocks [25] similar to Vertex-2 [26].

To facilitate the design and evaluation using 3D FPGAs, new CAD tools for 3D FPGA integration are needed. A CAD tool for an FPGA accepts an RTL description of a circuit and an architectural description of the FPGA for generating a configuration bitstream that can be loaded into the FPGA. This complex problems of mapping a circuit to an FPGA is broken down into a sequence of sub-problems or stages as illustrated in Fig. 5.4. Each of these problems can be reduced to some classical theoretical problems that have been extensively researched. Each stages has its own optimization goal which is related to the ultimate goal of satisfying area and performance requirements for the circuit when mapped to the FPGA. Several CAD tools have been proposed earlier for both 2D FPGAs and more recently most of them are extended for 3D FPGAs and all of them have almost similar tool flow which will be detailed in this section.

A graph-based 3D FPGA placement and routing algorithm presented in [19]. A fast partitioning based placement tool for 3D FPGAs called TPR (Three dimensional

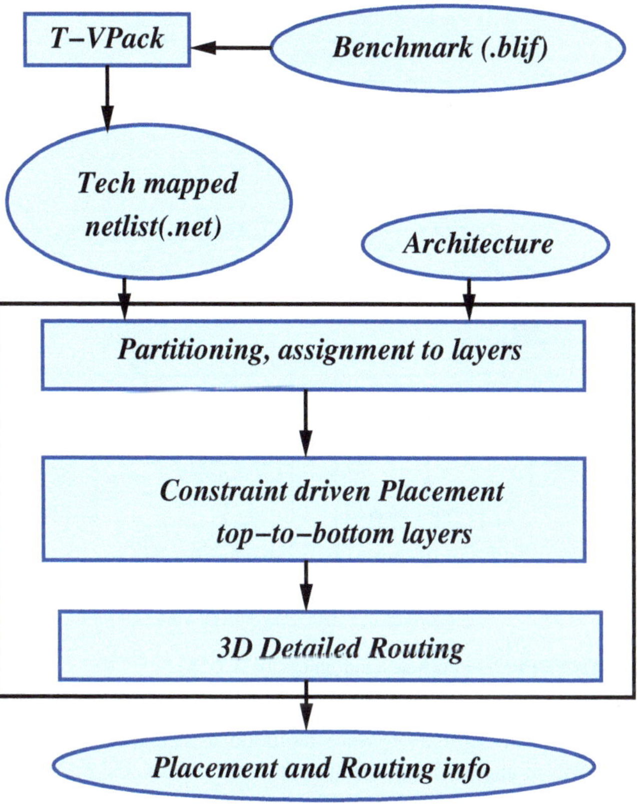

Fig. 5.5 TPR: Three-dimensional place and route tool for 3D FPGA, illustration of the 3FD FPGA CAD flow

Place and Route) is presented in [11, 27, 28]. The tool effectively investigate the impact of 3D integration on path delay of FPGAs in addition to interconnect wire-length. The philosophy of the tool closely follow that of its 2D counterpart VPR (versatile place and route) [29]. The flow of the TPR placement and routing CAD flow is shown in Fig. 5.5. Nevertheless the TPR router is not timing driven as well as all SBs are assumed to be 3D-SBs and they also assumed that the number of inter-tier vias (TSVs) is the same as the horizontal channel width. In today's technology, especially if we stack more than two active layers, the vias are much thicker than the horizontal wires (for example, 1–0.1 μm), which makes such assumption impractical.

A fully-fledged 3D FPGA design framework, called 3D-MEANDER for the design and exploration of 3D Mesh-based FPGAs presented in [10, 30]. This methodology provides the capability to analyze the impact of different deployment strategy for 3D-SBs in multi-tier Mesh-based FPGAs. This design methodology is a 3D extended version VPR, in which new CAD tools for partitioning, placement

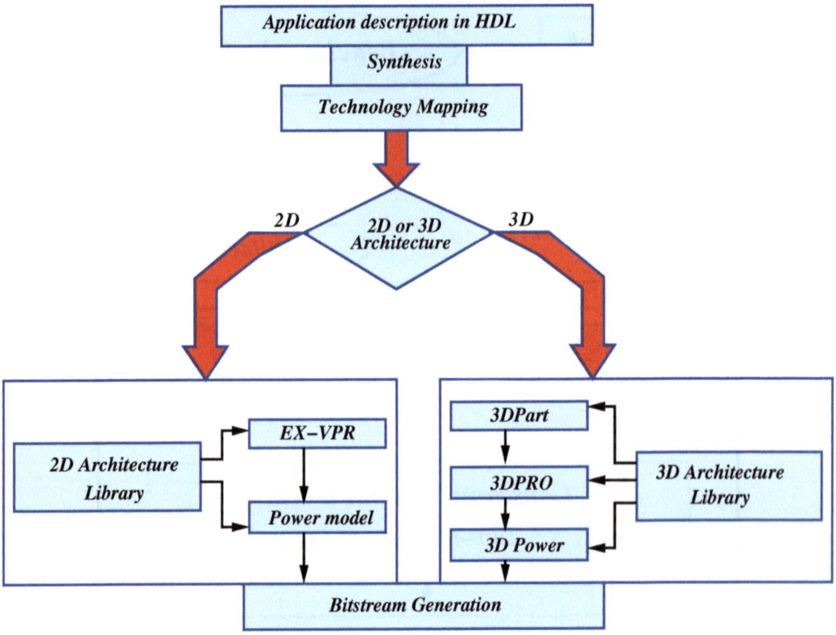

Fig. 5.6 The MEANDER 3D FPGA design framework [10, 30]

and routing and power estimation were included as illustrated in Fig. 5.6. It support the evaluation of alternative interconnect architectures and proposes a family of 3D Mesh-based FPGA interconnect architectures in which 2D-SBs and 3D-SBs are intermittently used in certain regular spatial patterns as illustrated in Fig. 5.7. The combination of 2D and 3D SBs may result in design and manufacturing issues. The 3D SBs are lager in size compared to 2D SBs due to increased number of vertical channels and support devices. A major drawback this methodology is that the number of available TSVs within 3D-SBs is assumed to be fixed and the design methodology does not the ability to investigate the impact of different numbers of TSVs in a 3D-SB.

Although several researchers have proposed 3D FPGAs, the detailed routing architecture of a 3D FPGA remains largely unexplored. In an FPGA the routing tracks and programmable switches constitute the routing channel. Channel width refers to the number of tracks in the channel. The LBs connect the channel through the connection boxes. The routing wires connect among themselves through the switch boxes. Switch-box topology refers to the connectivity provided by the switch box. Researches have explored several typologies [31, 32]. The subset (also called disjoint) topology, used in Xlinx XC4000 FPGAs, connect all tracks and net uses the same track number for its route as illustrated in Fig. 5.8. A universal 2D switch block is designed and presented in [33], which used track count as the sole metric of quality. Universal topology provides more flexibility than disjoint as illustrated in Fig. 5.8. It facilitates connectivity for all possible global routes of two-terminal nets.

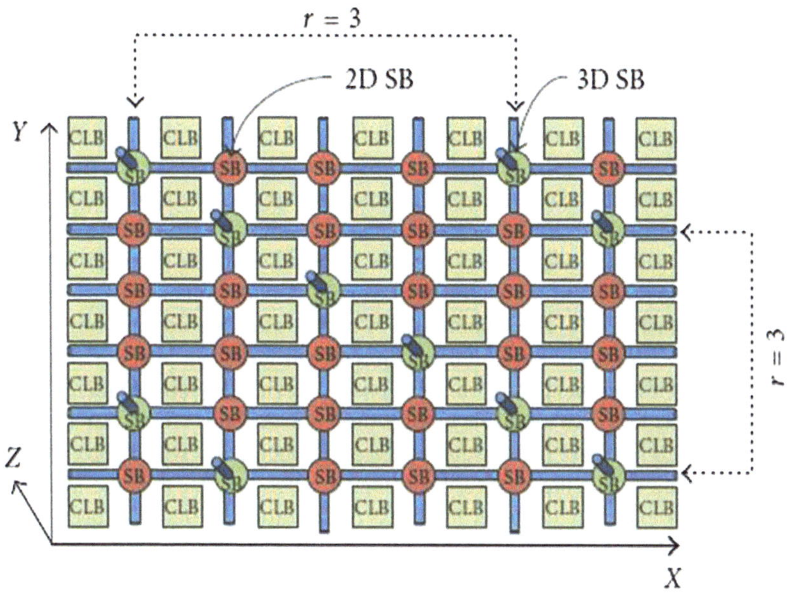

Fig. 5.7 The MEANDER 3D FPGA 2D and 3D SBs interconnect architecture [10, 30]

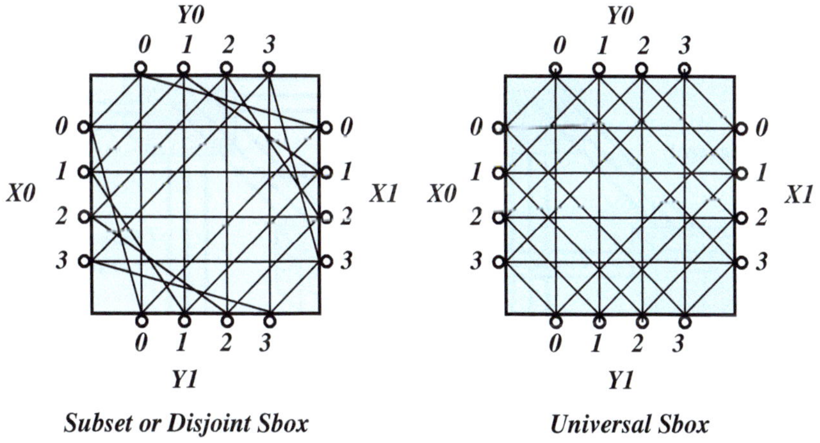

Fig. 5.8 Disjoint and universal 2D switch boxes, X_0,X_1,Y_0,Y_1 mark their sides

5.4 3D FPGA Interconnect Switch

The flexibility factor F_s of a switch box refers to the number of wires to which each incoming wire can connect. Previous studies have shown that for a 2D FPGA, an F_s of 3 provides good routability [34]. In such SBs, a track connects to one track on each of the other sides of SB. Disjoint and Universal topologies are examples of such SBs as illustrated in Fig. 5.8. In the case of island style Mesh-based FPGA, the 2D SBs are extended to 3D by adding two more faces, which contain terminal for vertical interconnection (Z direction), one for the layer above and another for the layer below as illustrated in Fig. 5.9. The channels in X and Y direction contain same number of tracks, however the channels in the Z direction differs from X and Y direction in its width, which is influenced by the via density provided by the 3D manufacturing technology. The length of channels in the Z direction depends on the wafer thickness as we discussed in Chap. 2, the wafer used in 3D integration will undergo thinning process and due to this the length of via can be much smaller than the average 2D wirelength, for example the wafer thickness of Tezzaron 3D manufacturing process is 12 μm, which includes 6 μm device layer (FEOL) and another 6 μm metal layer (BEOL) [35]. In effect, since the vertical vias will be fewer than horizontal wires, the two vertical faces will contain fewer terminal than the other four sides.

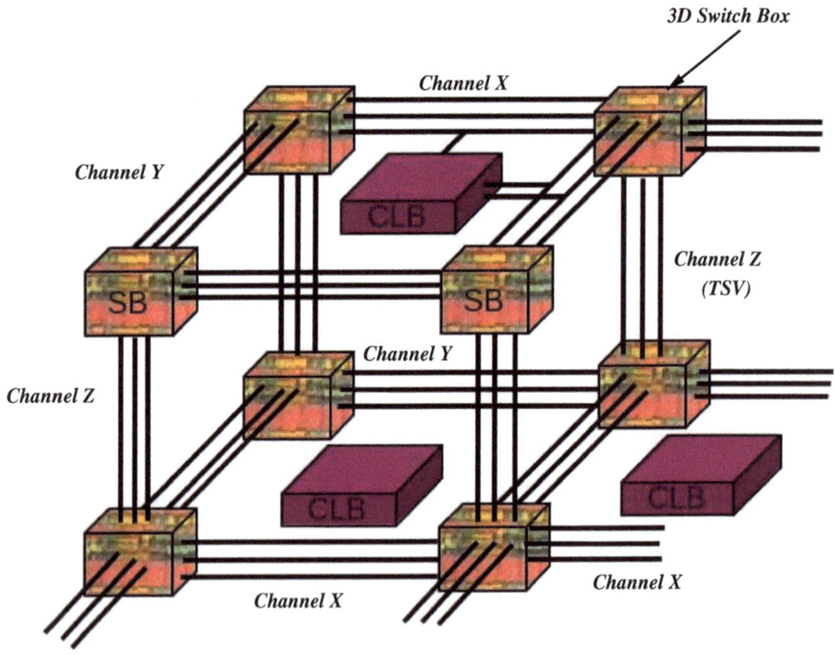

Fig. 5.9 3D Switch box with channel in X, Y ans Z direction

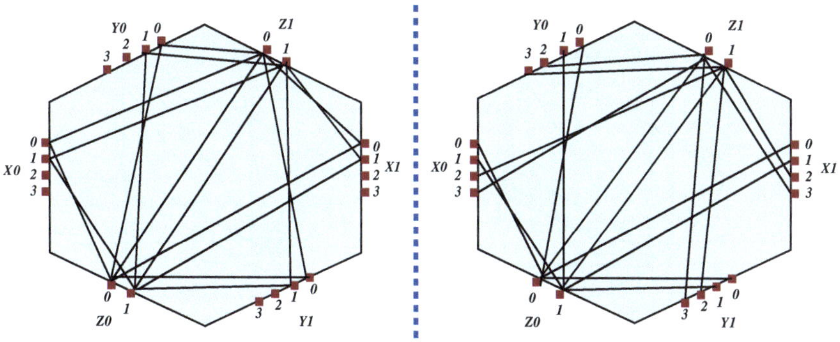

3D Disjoint Switch Box

Z0,0	X0,0	Y0,0	X1,0	Y1,0	Z1,0
Z0,1	X0,1	Y0,1	X1,1	Y1,1	Z1,1
Z1,0	X0,0	Y0,0	X1,0	Y1,0	Z0,0
Z1,1	X0,1	Y0,1	X1,1	Y1,1	Z0,1

3D Disjoint–split Switch Box

Z0,0	X0,0	Y0,0	X1,0	Y1,0	Z1,0
Z0,1	X0,1	Y0,1	X1,1	Y1,1	Z1,1
Z1,0	X0,3	Y0,3	X1,3	Y1,3	Z0,0
Z1,1	X0,2	Y0,2	X1,2	Y1,2	Z0,1

Fig. 5.10 3D Disjoint and disjoint-split switch box with channel in X=Y=4 an Z=2

The 3D SBs can be represented as a cube , where each face of the cube represent one of the direction. Nevertheless for ease of illustration, the 3D SB is represented as hexagon, where each side represent a direction: North (Y_0), South (Y_1), East (X_1), West (X_0), top (Z_1), and bottom (Z_0). To understand how the vertical connections are placed, we only show the connections to the vertical faces Z_0 and Z_1. For all SBs, the horizontal wires in *X and Y* direction uses either disjoint or universal connections among themselves. These connections are illustrated in Fig. 5.8. Figure 5.10 shows the 3D SB with disjoint topology. For clear understand of vertical connections, we do not shows the horizontal connections. The first SB in Fig. 5.10 is an extension of 2D disjoint SB. This SB connects the same track numbers to all sides. Consequently, the entire fabric gets divided into disjoint subsets, and a net uses the same track number for its entire route. The disjoint SB used in this study has two connections in its vertical faces and 4 connections in the horizontal direction, but only first two of horizontal wires connect to the vertical vias. While these wires have a flexibility of 5, in which 3 connections to the other horizontal directions and two to the vertical vias, which make the flexibility in horizontal direction 3. Apart from decreasing routing flexibility, this arrangement also results in a difference in capacitive load of the horizontal wires: larger capacitive load for the wires which share horizontal and vertical via connections and small for those wires with connections only in horizontal direction. The *disjoint-split* is modified version of disjoint SB to evenly distribute the capacitive load on the horizontal tracks. The difference is, we use the first 2 horizontal tracks to connect to via going above and the other two horizontal tracks connects the vias going below. This implies that now there are twice as many horizontal wires that connect to vertical vias. Therefor, if nets do not fan-out at the SB, then this style of connections provides greater flexibility to vertical directions. However, the

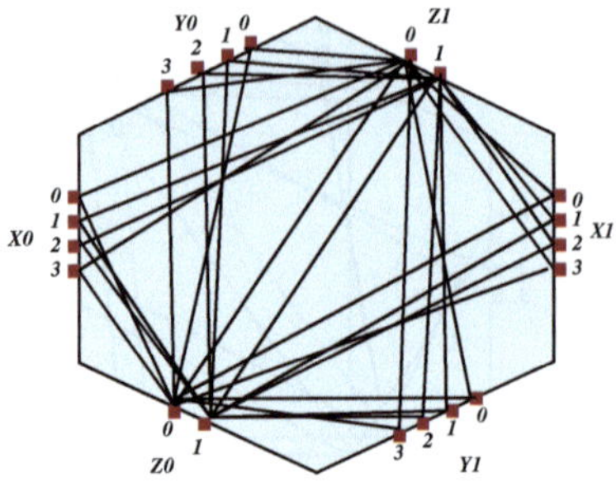

3D Disjoint–more Switch Box

Z0,0	X0,0	Y0,0	X1,0	Y1,0	Z1,0	X0,3	Y0,3	X1,3	Y1,3
Z0,1	X0,1	Y0,1	X1,1	Y1,1	Z1,1	X0,2	Y0,2	X1,2	Y1,2
Z1,0	X0,3	Y0,3	X1,3	Y1,3	Z0,0	X0,0	Y0,0	X1,0	Y1,0
Z1,1	X0,2	Y0,2	X1,2	Y1,2	Z0,1	X0,1	Y0,1	X1,1	Y1,1

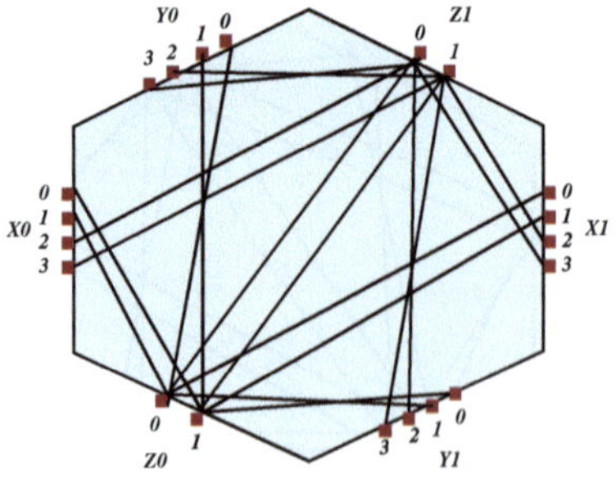

3D Disjoint–twist Switch Box

Z0,0	X0,1	Y0,0	X1,0	Y1,1	Z1,0
Z0,1	X0,0	Y0,1	X1,1	Y1,0	Z1,1
Z1,0	X0,2	Y0,3	X1,3	Y1,2	Z0,0
Z1,1	X0,3	Y0,2	X1,2	Y1,3	Z0,1

Fig. 5.11 3D Disjoint-more and disjoint-twist switch box with channel in X=Y=4 an Z=2

disadvantage of this SB is, if a net needs to fan-out to top as well as bottom vias, then it needs to use two horizontal tracks compared to one 3D disjoint SB.

The disjoint-split SB, although more flexible than disjoint SB, suffers from the disjoint property of the disjoint SB: the entire routing fabric is divided into disjoint subsets and a net can only use one of those subsets. In order to improve upon this issue, the disjoint-split modified the connections to vertical faces as shown in Fig. 5.11. With this change, the terminal *Z0,0* connects to track 1 one the side X0, but track 0 on side X1. This allows the net to switch tracks at the SBs. This type of connection topology is called *disjoint-twist* SB. The main objective of the disjoint-twist SB is to improve the flexibility in the vertical direction. Another way to achieve this is by adding more switches to the vertical direction and this approach is called *disjoint-more* SB as illustrated in Fig. 5.11. The extra switches have twofold effect. On the one hand, they improve the flexibility in the vertical direction, and on the other hand, they increase the area of SB and the capacitive loads on the wires.

The 3D SBs presented in Fig. 5.12 use *universal* connections among the horizontal wires. The vertical connections in the *Universal-twist* SB are identical to the *disjoint-twist* SB as illustrated in Figs. 5.11 and 5.12. However, due to universal connections among the horizontal wires, it provides greater flexibility. The *Universal-more* further increase the flexibility by adding more switches to the vertical faces. These extra switches improve the flexibility in the vertical direction, but also increase the area of the SB and the capacitive load on the wires [36].

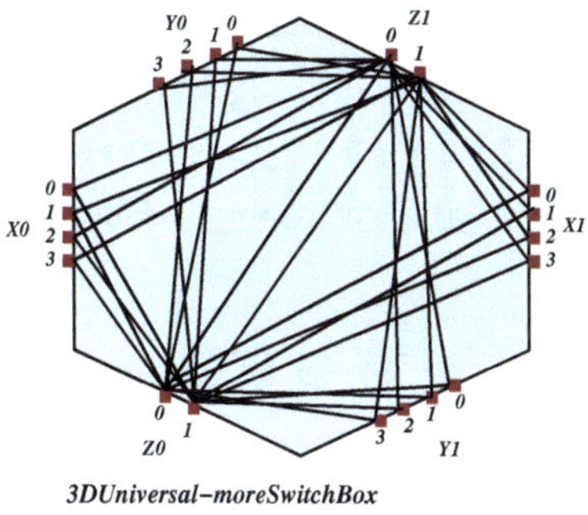

Z0,0	X0,0	Y0,0	X1,0	Y1,0	Z1,0	X0,3	Y0,2	X1,2	Y1,3
Z0,1	X0,1	Y0,1	X1,1	Y1,1	Z1,1	X0,2	Y0,3	X1,3	Y1,2
Z1,0	X0,2	Y0,3	X1,3	Y1,2	Z0,0	X0,0	Y0,1	X1,1	Y1,0
Z1,1	X0,3	Y0,2	X1,2	Y1,3	Z0,1	X0,1	Y0,0	X1,0	Y1,1

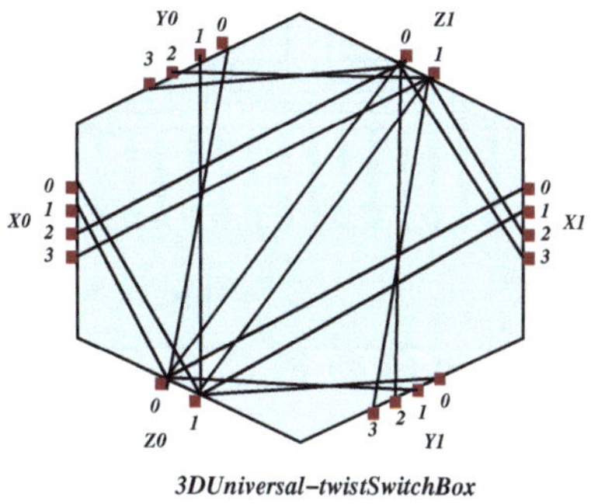

Z0,0	X0,1	Y0,0	X1,0	Y1,1	Z1,0
Z0,1	X0,0	Y0,1	X1,1	Y1,0	Z1,1
Z1,0	X0,2	Y0,3	X1,3	Y1,2	Z0,0
Z1,1	X0,3	Y0,2	X1,2	Y1,3	Z0,1

Fig. 5.12 3D Universal-more and universal-twist switch box with channel in X=Y=4 an Z=2

5.5 2.5D Integration: High Density Multi-FPGAs

Interposer-based silicon die integration is called 2.5D integration. An interposer-based stacking gained popularity last year because of several attractive applications as well as its technological feasibility. It might not be sufficiently effective for other applications, however, such as the memory on logic or the logic splitting application, where the logic is split between two or more dies that are then put on top of each other for shorter interconnections. Also memory can be split in such a way that read-write logic is on one chip while the cells are on the other. A true 3D stacking is needed for maximum performance in those applications. Interposer-based multi-FPGA systems are composed of multiple FPGA dice, which are connected through a silicon interposer as illustrated in Fig. 5.13. The interposer is fabricated with an older technology than that used for the dies, and links between dice include both a micro-bump on each FPGA die and a metal wire on the interposer [37]. This results in a reduced connectivity between dice and increased delay for connections between dice, as compared to the total routing connectivity and the delay one can achieve within a single die. The silicon interposer FPGA technology is interesting because it enables the creation of large FPGAs composed of small dies and also of very large FPGAs, with higher logic capacity than one cannot achieve with a single die. Such FPGA systems are sometimes called 2.5D FPGAs, since they make use of vertical stacking of dies on an interposer to enable higher integration levels. Being able to make large FPGAs with multiple smaller dies is particularly interesting at the beginning of a new manufacturing process, when defect densities are high. In such a case, good-die yield drops dramatically as the die size increases, and this drastically impacts the availability of large FPGAs early in the process lifetime. Effectively the interposer-based FPGAs allow the creation of chips larger than a single die, making

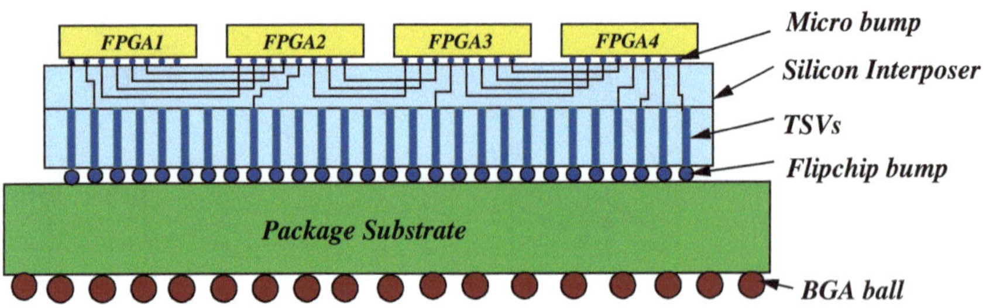

Fig. 5.13 The structure of an interposer-based 3D FPGA. The structure has a silicon interposer between FPGA dies and package

a *More than Moore* improvement on the size and number of logic elements possible, and with chips combined far more tightly and with more connectivity than if they were on separate boards connected through conventional means.

Xilinx's approch to interposer-based 2.5D FPGA is presented in [9]. They describe the physical characteristics of their implementation, including the bump pitch and estimates of the amount of die-to-die connectivity and the die-to-die delay. The improvement in the number of logic elements of 2.5D FPGAs over conventional ones is very significant. Xilinxs largest interposer-based FPGA, the *Virtex-7 XC7V2000T*, has 4 dies (which Xilinx calls Super Logic Regions) and 1.954 million logic elements [38], while the largest non-interposer Virtex-7 die (the XC7VX980T), has 979 K logic elements and Alteras largest FPGA, the Stratix V 5SGXBB, has 952 k logic elements [39]. Even though all these FPGAs use a 28 nm process, silicon interposer technology allows the creation of FPGAs with twice the resources possible on even an extremely large single die. Another major advantage of interposer-based FPGAs comes at the beginning of a new manufacturing process, when the defect density is high [37]. Bigger dice suffer a much reduced yield compared to smaller dice, and this greatly affects the supply and cost of top-of-the-line FPGAs to early adopters. To illustrate this impact, consider a new process in which the defect density is $1/\text{cm}^2$, which is a reasonable value early in the process life-cycle, and the die area is $6\,\text{cm}^2$, which roughly matches the size of the largest member of a high-end FPGA family such as Virtex 7. Using the Poisson Yield Model [6], the yield is only 0.25 % of die. If instead the chip is composed of four $1.5\,\text{cm}^2$ dies, the yield is 22 %. This means that a 12 in. silicon wafer with $730\,\text{cm}^2$ of area would produce on average 0.3 working $6\,\text{cm}^2$ dies, while the same wafer would produce on average 107 working $1.5\,\text{cm}^2$ dies. Therefore, as a $6\,\text{cm}^2$ chip would be composed of four $1.5\,\text{cm}^2$ dies, the wafer would yield 26.75 systems on average, as the *assembly yield* of placing these four die on an interposer is very high [9]. Hence the number of interposer-based FPGAs created from the same silicon wafer would be almost 100X greater than a monolithic FPGA of the same size.

When the process matures and the defect density decreases this advantage drops significantly. Consider a mature process with defect density of $0.1/\text{cm}^2$. The yield for a 6 cm^2 die is 55 % and the yield for a 1.5 cm^2 die is 86 %. Hence the number of single die FPGAs created from a 730 cm^2 silicon wafer would be 66.9 and the number of interposer-based FPGAs created would be 104.6. While the interposer-based FPGA still has a yield advantage it might not lead to a major cost advantage, particularly when the cost of the interposer and assembling the die to it are included. For the large, state-of-the-art FPGAs that are built early in a process cycle and heavily used for prototyping, however, there is clearly a compelling cost advantage to an interposer-based solution. However the main difference between 2.5D interposer-based technology and 3D integration is that 2.5D technology uses a passive silicon interposer, while true 3D technology uses active silicon interposer. Second point is, interposer-based technology can be manufactured in Front-end (FEOL) or Back-end (BEOL). The Front-end interposer technology may increase the cost. But for true 3D integration Front-end manufacturing is preferred. Last point is the 2.5D and 3D technology are currently evolving and the are to be considered as evolutionary parallel technology, but 2.5D technology does not necessarily evolve to 3D Technology.

5.5.1 Industrial 2.5D Virtex-7 Interposer-Based FPGAs

The Xilinx 2.5D FPGAs from the Virtex-7 family are currently the only commercially available silicon interposer-based FPGAs. The lateral view of interposer-based FPGA [8, 9, 40] presented in Fig. 5.14. Inside the 2.5D package, a hetero-

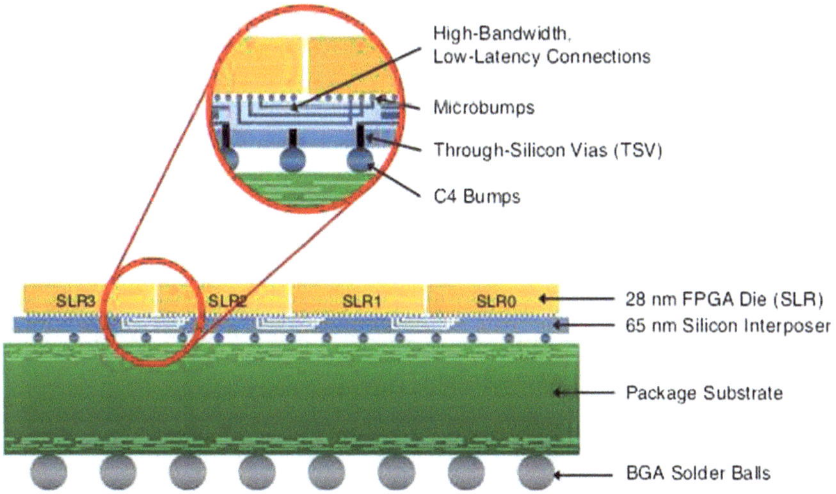

Fig. 5.14 Lateral view of an interposer-based FPGA [9, 40]. The FPGA dice are at the *top*, and are connected to the silicon interposer through microbumps. The interposer is then connected to the substrate through C4 bumps

geneous IC stack delivers up to 2.78 Tb/s transceiver bandwidth. The resulting bandwidth is approximately 3 times achievable in a monolithic FPGA chip. The 2.5D FPGA is mounted on passive interposer with TSVs, the heterogeneous IC stack comprises FPGA dies with 13.1 Gb/s transceivers and dedicated analog ICs with 28 Gb/s transceivers. The $XC7VH580T$ [8] is the first commercial FPGA built with heterogeneous interposer-based interconnect. The device consists of a passive silicon interposer and three active die: an 8×28 Gb/s transceiver IC and 2 FPGA ICs. Three additional products are derived from these building blocks: The $XC7VH290T$ with one FPGA die and $XC7VH870T$ consisted of 3 FPGA dies, one 16×28 Gb/s transceivers and 72×13.1 Gb/s transceivers which allow the delivery of 2.78 Tb/s. The *XC7V2000T* is composed of four identical dice arranged such that the vertical routing crosses between the dice. Each horizontal edge of each die has 280 groups of 48 length-12 wires crossing the interposer, which sums to a total of 13,440 wires between dice. There are also 40 clock wires crossing the interposer. The average number of wires per vertical channel of this FPGA is 210 and there are approximately 280 vertical channels on the FPGA, resulting in approximately 58,800 vertical wires crossing a horizontal cut-line within a die. Hence the number of wires which cross the interposer is about 23 % of the total number of within-die vertical wires. The TSV interposer wafers are manufactured by etching vias through silicon wafers and filling the vias withe a conductive metal, typically copper. The TSVs in an interposer in manufactured using via-first/via-middle flow since it offers the greatest benefit of interconnect density. The TSVs are typically 10–20 μm in diameter and 50–100 μm deep. The walls of the TSVs are lined with S_iO_2 dielectric. Then a diffusion barrier and a copper seed layer are formed. The via hole is filled with copper through electrochemical deposition. The local interconnect wires M1-Mx (Type I and Type II signals) are formed on top of the interposer using standard backend (BOEL) fab process. Interposer top side is coated with passivation and micro-bump pads are formed. The interposer wafer in thin down to TSVs to expose the TSV from the bottom side.

The 28 nm dies are connected to the 65 nm silicon interposer through microbumps with a 45 μm pitch. Hence the area occupied by microbumps at one edge of one die is $13,440(45\,\mu\text{m})^2 = 27\,\text{mm}^2$. If we assume each die is 7 12 mm, as presented in [9], the bumps have to be spread out near the edge and need to go as far as 2.25 mm away from the edge of the die. This greater distance from the border increaes the length of the inter-die connections, and along with the presence of the micro-bumps and their capacitance, leads to an increased delay for these crossing wires vs. that of a typical on-die routing wire. The latency to cross the interposer is approximately 1 nS as stated in [9]. For comparison, a typical medium length 28 nm FPGA routing wire (e.g. spanning four logic blocks) has a delay of approximately 125 pS, while a longer wire (e.g. spanning 12 logic blocks) has a delay of approximately 250 pS. Overall, these interposer-based FPGAs have increased delay and reduced connectivity between dies, with approximately 23 % of the usual number of vertical wires crossing between dies and approximately 1 nS of increased delay to cross the interposer.

5.6 Development of 3D Tree-Based FPGA CAD Tools

Three-dimensional FPGA architectures have been widely investigated over last few years [10, 11, 18, 30, 36]. A major aspect of 3D-FPGA architecture research is the development of CAD tools for mapping application to FPGA and to design and manufacture 3D FPGA based systems. It is well established fact that the quality of a high density multi-tier or multi-chip FPGA based implementation is largely determined by the effectiveness of accompanying CAD tools [11, 30, 41]. Benefits of an otherwise will designed, high density and faster FPGA architecture might be impaired if the CAD tools cannot take the advantage of new features and characteristics of the new FPGA architecture. Thus the design and development 3D FPGA CAD algorithm research and exploration is essential to bring the high logic density and fast 3D FPGAs into today's semiconductor market. We developed 3D FPGA CAD flow to specifically design and implement 3D Tree-based and Mesh of Tree-based FPGA architectures. The 3D CAD flow for physical design and architecture exploration and optimization are developed in three different stages as illustrated in Fig. 5.15. The first section is the 3D physical design flow for Tree-based FPGA architecture where we combined many industry standard design tools and additional home grown tools to support 3D stacking and design verification. The Tree-based FPGA architecture exploration and optimization methodologies are developed with same flexibility and capability to implement netlists with equivalent congestion like the Mesh-based FPGA architectures. Thus architecture evaluation and optimization must be based on benchmark circuits implementation.

5.6.1 3D FPGA Physical Design Tools

The physical design process begins with the RTL description of Tree-based FPGA generated using VHDL code generator as illustrated in Fig. 5.15. We then used cadence design compiler to compile VHDL into structural Verilog for each die. The compiled Verilog is then input into Cadence Encounter to perform semi-automated physical design steps. We used a divide and stack design methodology to implement multi-tier 3D Tree-based FPGA. Based on the design and speed requirements, we partitioned the design into two or more tiers and later merged them to conduct the (DRC/LVS) 3D design verification process. The design tool augmented to test different 3D stacking methodologies. We used both Face-to-Face (F2F) and Face-2-Back (F2B) stacking methodology provided by Tezzaron's 3D design platform using via first TSV process. The Tezzaron's 3D stacking kit support two types of TSV structures: Supper-Contact and Super-Via. In our design we used Supper-Contact, using tungsten via fill. The place and route for multi-tier 3D Tree-based FPGA is performed using Cadence Encounter. A well advanced 3D thermal model also integrated along with the 3D physical design flow to check the thermal profile of the 3D chip. A detailed 3D thermal analysis of Tree-based FPGA is presented in Chap. 7. After

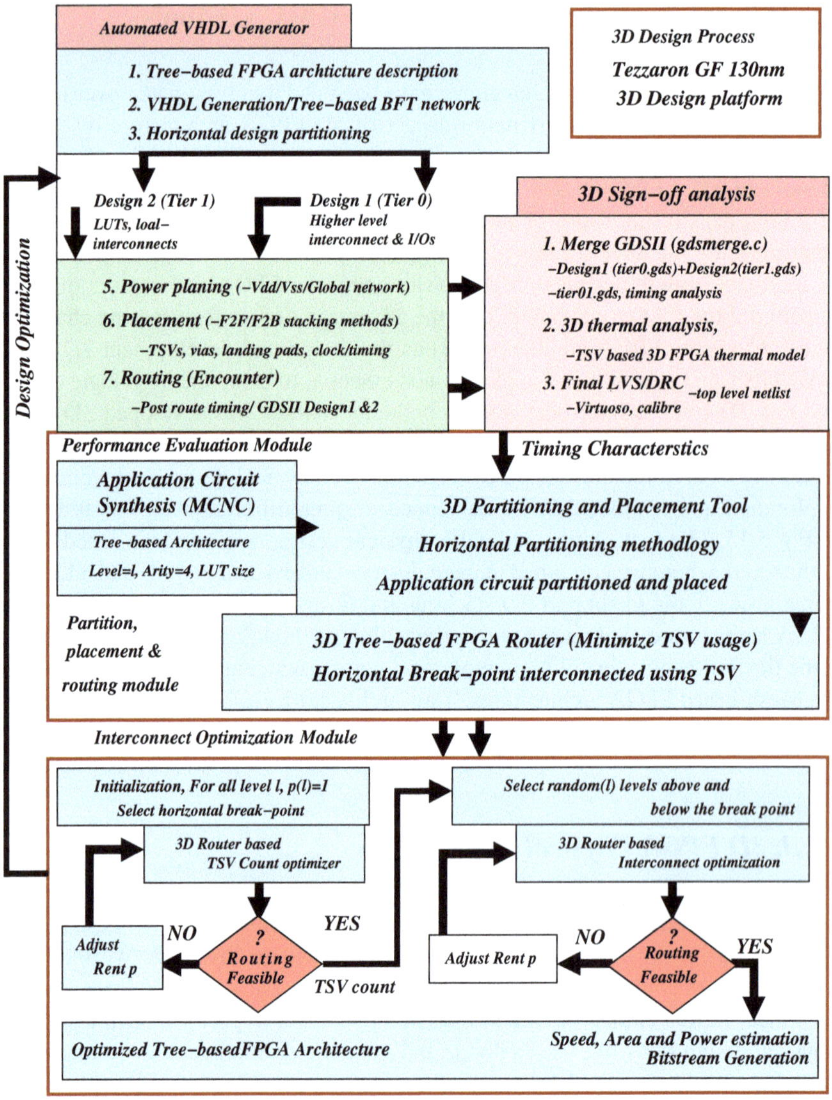

Fig. 5.15 Design and exploration software flow: 3D physical design and architecture optimization flow for design and implementation of horizontally partitioned 3D Tree-based FPGA

completing DRC/LVS individual designs, we used the GDS merger tool to integrate the different tiers of the 3D FPGA chip. We then used Calibre-LVS compares this merged GDS file with top level schematic. A detailed design analysis and experimentation of two-tier 3D Tree-based FPGA presented in Chap. 8.

5.6.2 3D FPGA Architecture Exploration and Optimization

The first part of the architecture exploration flow deals with synthesis and conversion of application netlist to .net format, while remaining sections deals with architecture exploration. The method used for synthesis and conversion of application is presented in Chap. 4, Sect. 4.7.1. We use a top-down recursive partitioning and clustering approach. The LBs, HBs and IOs are partitioned into different clusters. The aim is to reduce external communications and to collect highly connected cells into the same cluster. First, we construct the top level clusters, then each cluster is partitioned into sub-clusters, until the bottom level of the architecture is reached. Then during the placement phase, each cluster is assigned to a random position inside its owner cluster. The partitioning in each level consists of three phases. First we run a multilevel coarsening phase where the size of hypergraph is successively decreased using the first choice algorithm [42]. Then k-way partitioning of the smaller hypergraph in computed such that the balancing constraint is satisfied. After that we run the un-coarsening phase where the partitioning is successively refined using using FM algorithm [43], as it is projected in the larger hypergraphs. The objective of the refinement is to minimize the hyperedge-cut, which is the total number of hyperedges that span multiple partitions. During the refinement, a block with highest gain is moved from one partitioned to another and then its locked and it is not allowed to move during the remaining refinement phase. After the block is moved, the gain of all associated blocks are recomputed and this process continues until all the blocks are locked. At the end, total cost is compared to that of the previous solution and the algorithm is terminated when it fails to improve during refinement. Since the structure of Tree is maintained in our fully connected two-tier 3D FPGA, the 3D design break-point will not play any role in application partitioning and placement process. However it is used during architecture optimization process. Figure 5.15 presents the block level representation of Tree-based FPGA architecture exploration platform.

The design netlist to be mapped are obtained in .NET format. The LUTs, I/Os and Hard-blocks (HBs) are first partitioned into different clusters in such a way the inter-cluster communication is minimized while considering the horizontal or vertical partition of programmable BFT networks. After completing the partition, the placement file is generated. It contains positions of different blocks in the architecture. This placement file along with netlist file and Tree-based FPGA architecture files are then passed to 3D router, which is responsible for routing the application netlist. Router is based on PathFinder [44–46] routing algorithm that uses an iterative, negotiation-based approach to successfully route all nets in an application netlist. In order to rout all all nets of the netlist, routing resources of the multi-tier interconnect structure are first assigned to the respective blocks of the netlist that are placed on the architecture. These routing resources are modeled as directed graph abstraction *G(V,E)*. In this graph the set of vertices *V* represent the in/out pins of different blocks and routing wires in the interconnect structure and an edge *E* between two vertices, represents a potential connection between to vertices. The edges interconnect break-point nets are considered as 3D nets which uses TSV to interconnect between vertices on tier 0 and 1. The 3D timing analyzer generates direct acyclic timing graph of the routed circuit of the multi-tier 3D chip to evaluate the critical path delay. Based on routing result, the different sub-paths are identified and each edge is annotated with delay of corresponding sub-path. The edges interconnect vertices from tier 1 to 0 of the 3D Tree-based heterogeneous FPGA annotate corresponding TSV delay. The timing graph specifies edges interconnect tier 0 and 1 as 3D nets. To optimize the TSV count and routing resources, a Rent-based wire length distribution model implemented using 3D router is used. After completing the TSV count and architecture optimization, the router estimates the critical path delay, TSV count, area and power consumption of the optimized 2 tier 3D stacked heterogeneous Tree-based FPGA.

5.7 Summary

Many researchers and industrial partners demonstrated that 3D FPGA can provide significant advantages over 2D by reducing the interconnect area and the total area-delay product. The five-tier 3D Mesh-based FPGA presented in [36] shows area-delay product improved by 36 % compared to 2D FPGA of identical logic density. We also reviewed several types of 2D and 3D SBs used to design 3D FPGAs. The result analysis shows the area-delay product depend heavily on the SB topology. The 3D experimental analysis shows the *universal switch box* is a better choice for process technology scaling and 3D implementation as it can accommodate more number of vertical vias. A 3D design framework for 3D Mesh-based FPGAs using asymmetrical interconnect fabric is presented in [30]. The 3D FPGA design shows 26 % reduction in total wire-length and 38 % reduction in path delay. Nevertheless the design methods reported in [30] are not practical because it requires complete custom design due the heterogeneous nature of 2D and 3D SB distribution inside 3D FPGA array. 3D Mesh-based FPGA implementation using 3D monolithic technology presented in [12, 24]. The results analysis shows the 3D monolithic FPGA improve the logic density by 3.2 times and 1.7 times lower critical path delay and power consumption compared to 2D monolithic FPGA. The results are encouraging for 3D monolithic FPGA, however the process integration seems not going to be ready in near future for such designs.

References

1. S.M. Trimberger, *Field Programmable Gate Array Technology, Norwell* (Kulwer, MA, 1994)
2. A. DeHon, Reconfigurable Architectures for General-Purpose Computing. Ph.D. dissertation, Department of Electrical Engineering and Computer Science, Massachusetts Institute of Technology, 1996
3. I. Kuon, J. Rose, Measuring the gap between FPGAs and ASICs. IEEE Trans. Comput. Aided Des. Integr. Circ. Syst. 26(2), 203–215 (2007), http://dx.doi.org/10.1109/TCAD.2006.884574 (IEEE Council on Electronic Design Automation)

4. L. Shang, A.S. Kaaviani, K. Bathala, Dynamic power consumption in vertex-II FPGA family, in *Proceeding of ACM/SIGDA International Conference on FPGAs*, pp. 157–164 (2002)
5. E. Ahmed, J. Rose, The effect of LUT and cluster size on deep-submicron FPGA performance and density. IEEE Trans. Very Large Scale Integr. (VLSI) Syst. **12**(3), 288–298 (2004)
6. A. Rahman, R. Reif, System level performance evaluation of three-dimensional integrated circuits. IEEE Trans. Very Large Scale (VLSI) Syst. **8**, 671–678 (2000)
7. A. Rahman, H. Shi, Z. Li, D. Ibbotson, S. Ramaswami, in *Design and Manufacturing Enablement for Three-Dimensional (3D) Integrated Circuits (ICs)*. CICC, pp. 1–8 (2012)
8. L. Madden, E. Wu, N. Kim, B. Banijamali, K. Abugharbieh, S. Ramalingam, X. Wu, Advancing high performance heterogeneous integration through die stacking, in *ESSCIRC*, pp. 18–24 (2012)
9. R. Chaware, K. Nagarajan, S. Ramalingam, Assembly and reliability challenges in 3D integration of 28nm FPGA die on a large high density 65nm passive interposer, in *IEEE International Conference on ECTC*, pp. 279–283 (2012)
10. K. Siozios, A. Bartzas, D. Soudris. Architecture level exploration of alternative schmes targeting 3D FPGAs: a software supported methodology. Int. J. Reconfig. Comput. **2008**, 2008
11. C. Ababei, P. Maidee, K. Bazargan, Exploring potential benefits of 3D FPGA integration, in *Field Programmable Logic and Application*, vol. 3203 (Springer, Berlin, Germany, 2004), pp. 874–880
12. M. Lin, A. El Gamal, Y.-C. Lu, S. Wong, Performance benefits of monolithically stacked 3D FPGA, in *Proceedings of the 2006 ACM/SIGDA 14th International Symposium on Field Programmable Gate Arrays*, Monterey, California, USA, pp. 113–122, 22–24 Feb 2006
13. S. Wong, A. El-Gamal, P. Griffin, Y. Nishi, F. Pease, J. Plummer, Monolithic 3D integrated circuits, in *International Symposium on VLSI Technology, Systems and Applications, 2007* VLSI-TSA 2007
14. P. Batude, M. Vinet, A. Pouydebasque, C. Le Royer, B. Previtali, C. Tabone, J.-M. Hartmann, L. Sanchez, L. Baud, V. Carron, A. Toffoli, F. Allain, V. Mazzocchi, D. Lafond, O. Thomas, O. Cueto, N. Bouzaida, D. Fleury, A. Amara, S. Deleonibus, O. Faynot, Advances in 3D CMOS sequential integration, in *IEEE International Electron Devices Meeting (IEDM)*, Baltimore, MD, Dec 2009
15. C. Liu, S.K. Lim, A Design tradeoff study with monolithic 3D integration, in *IEEE ISQED*, pp. 529–536 (2012)
16. M. Cao, T. Zhao, K.C. Swraswat, J.D. Plummer, A simpler EEPROM cell using polysilicon thin file transistors. IEEE Electron Device Lett. **15**(8), 304–306 (1994)
17. S. Kaeriyama, T. Sakamoto, H. Sunamura, M. Mizuno, H. Kawaura, T. Hasegawa, K. Terabe, T. Nakayama, M. Aono, A nonvolatile programmable solid-electrolyte nanometer switch. IEEE J. solid state circ. **40**(1), 168–176 (2005)
18. A. Rahman, S. Das, A. Chandrakasan, R. Reif, Wiring requirements and three-Dimensional integration of field programmable gate arrays, in *SLIP ACM*, March 2001
19. M.J. Alexander, P.J. Cohoon, J.L. Colflesh, J. Karro, G. Robins, Three-dimensional field-programmable gate arrays, in *Proceedings of the Eighth Annual IEEE International ASIC Conference and Exhibit*, Austin TX, pp. 253–256, Sept 1995
20. J.V. Campenhout, H.V. Marck, J. Depreitere, J. Dambre, Optoelectronic FPGAs. IEEE J. Sel. Top. Quantum Electron **5**(2), 306–315 (1999)
21. Miriam Leeser, Waleed M. Meleis, Mankuan M. Vai, Silviu Chiricescu, Xu Weidong, Paul M. Zavracky, Rothko: a three-dimensional FPGA. IEEE Des. Test **15**(1), 16–23 (1998)
22. Gaetano Borriello, Carl Ebeling, Scott A. Hauck, Steven Burns, The triptych FPGA architecture. IEEE Trans. Very Large Scale Integr. (VLSI) Syst. **3**(4), 491–501 (1995)
23. S. Chiricescu, M. Leeser, M. Michael Vai, Design and analysis of a dynamically reconfigurable three-dimensional FPGA. IEEE Trans. Very Large Scale Integr. (VLSI) Syst. **9**(1), 186–197 2001
24. O. Turkyilmaz, G. Cibrario, O. Rozeau, P. Batude, F. Clermidy, 3D FPGA using high-density interconnect monolithic integration, in *IEEE Design, Automation and Test in Europe Conference and Exhibition (DATE)*, March 2014

25. Y.-S. Kwon, P. Lajevardi, A.P. Chandrakasan, F. Honore, D.E. Troxel, A 3-D FPGA wire resource prediction model validated using a 3-D placement and routing tool, in *Proceedings of the 2005 International Workshop on System Level Interconnect Prediction*, San Francisco, California, USA, 02–03 April 2005
26. Vertex-5, Xilinx Inc., Vertex-5: Multi-platform FPGA, http://www.xilinx.com/products/silicon_solutions/fpga/vertex/vertex5/
27. C. Ababei, Y. Feng, B. Goplen, H. Mogal, T. Zhang, K. Bazargan, S. Sapatnekar, Placement and routing for 3D integrated circuits. IEEE Des. Test, **22**(6), 520–531 2005
28. C. Ababei, H. Mogal, K. Bazargan, Three-dimensional place and route for FPGAs. IEEE Trans. Comput. Aided Des. Integr. Circ. Syst. **25**(6), 1132–1140 (2006)
29. V. Betz, J. Rose, VPR: a new packing placement and routing tool for FPGA research. Int. Conf. Field Program. Logic Appl. **1997**, 213–222 (1997)
30. K. Siozios, Vasilis F. Pavlidis, D. Soudris. A novel framework for exploring 3-D FPGAs with heterogeneous interconnect fabric. ACM Trans. Reconfig. Technol. Syst. **5**(1), March 2012
31. S. Joseph, E. Wilton, J. Rose, Z. Vranesic, Architectures and Algorithms for Field-Programmable Gate Arrays with Embedded Memory. Ph.D. dissertation, Depatrment of Electrical and Computer Engineering, University of Toronto, Toronto, Ont., Canada, 1997
32. M.I. Masud, S. Joseph, E. Wilton, A New switch block for segmented FPGAs, in *Proceedings of the 9th International Workshop on Field-Programmable Logic and Applications*, pp. 274–281, 30 August–01 September 1999
33. G.M. Wu, M. Shyu, Y.W. Chang, Universal switch block from three-dimensional FPGA design, in *Proceeding of ACM/SIGDA International Symposium on Field Programmable Gate Arrays*, (1999)
34. J. Rose, S. Brown, Flexibility of interconnect structures for field programmable gate arrays. IEEE J. Solid State Circ. **26**, 277–282 (1991)
35. S. Gupta, M. Hilbert, S. Hong, R. Patti, *Techniques for Producing 3D ICs with High-Density Interconnect* (Tezzaron Semiconductor Naperville, IL, 2005)
36. A. Gayasen, V. Narayanan, M. Kandemir, A. Rahman, Designing a 3-D FPGA: switch box architecture and thermal issues. IEEE Trans. Very Large Scale Integr. (VLSI) Syst. **16**(7), 882–893 (2008)
37. K. Namhoon, D. Wu, D. Kim, A. Rahman, P. Wu, Interposer design optimization for high frequency signal transmission in passive and active interposer using through silicon via (TSV), in *IEEE Electronic Components and Technology Conference (ECTC)*, pp. 1160–1167 (2011)
38. Xilinx-7. 7 series FPGA overview (2013), www.xilinx.com
39. Altera. stratix V device overview (2013), www.altera.com
40. Xilinx stacked silicon interconnect technology delivers breakthrough FPGa capacity, bandwidth, and power efficiency (2012), www.xilinx.com
41. A.H. Pereira, V. Betz, CAD and routing for interposer based multi-FPGA systems, in *International Symposium on FPGAs*, CA, USA, 2014
42. N. Selvakkumaran, G. Karypis, Multi-objective hypergraph-partitioning algorithm for cut and maximum subdomain-degree minimization. IEEE Trans. Comput. Aided Des. Integr. Circ. **25**(3), 504–517 (2006)
43. C.M. Fiduccia, R.M. Mattheyeses, A liner-time heuristic for improving network partitions. ACM, Des. Autom. Conf. **7**, 175–181 (1982)
44. L. McMurchie, C. Ebeling, PathFinder: a negotiation based performance driven router for FPGAs. Int. Conf. Field Program. Gate Arrays FPGA **12**(2), 291–301 (1995)
45. Z. Marrakchi, H. Mrabet, H. Mehrez, Hierarchical FPGA clustering to improve routability, in *Ph.D. Research Conference in Microelectronics, PRIME*, 2005
46. Z. Marrakchi, H. Mrabet, U. Farooq, H. Mehrez, FPGA Interconnect topologies exploration. Int. J. Reconfig. Comput. **2009**, (2009)

Chapter 6
Three-Dimensional Tree-Based FPGA: Architecture Exploration Tools and Technologies

Abstract Modern FPGAs have become a viable alternative to cell-based design technology by providing re-configurable computing platforms with improved performance and higher density using 3D integration technology. While the re-configurability provides flexibility, FPGAs also lead to area and performance overhead in comparison to cell-based custom integrated circuits (ICs). Thus to combine the advantages of both FPGAs and custom ICs, modern 3D heterogeneous FPGAs emerged as an attractive solution for system-on-chip implementations. The modern FPGAs include design components such as digital signal processors, on chip memory blocks, multipliers, adders, and entire processors. In this chapter our primary focus is on teaching the development of 3D FPGA tools and technologies and the validation of architecture exploration tools and optimization methodologies by using custom designed 3D homogeneous and heterogeneous Tree-based FPGAs.

6.1 Introduction

According to the experimental and simulation results from the existing 3D FPGA CAD tools such as TPR and 3D MEANDER [1, 2] the SBs has been the most area consuming unit compared to other design elements in 2D FPGAs and this situation is becoming even worse in 3D FPGAs because the TSVs are located on 3D-SBs. Although the design and manufacturing engineers are trying to reduce TSV dimensions, the minimum feature size on the die is also shrinking. Therefore, the TSVs are expected to remain larger than wire dimensions in metal layers within the die [3]. Moreover it has been reported in [1, 2, 4] that the TSV utilization is actually quite low if the 3D-SBs are with full vertical connectivity in use. Many other recent research results also point out that the utilization of TSVs is actually very low in 3D Mesh-based FPGAs [2, 4, 5] with full vertical connectivity, which motivated us to explore Tree-based interconnect network, that can be optimized to achieve higher speed, reduced power consumption, area and increased logic density. As described in Chaps. 3 and 5, we explore a 3D Tree-based programmable interconnect to implement a fast and high density 3D FPGA. This chapter deals with validation and exploration of 3D homogeneous and heterogeneous Tree-based FPGA architecture and validation

V. Pangracious et al., *Three-Dimensional Design Methodologies for Tree-based FPGA Architecture*, Lecture Notes in Electrical Engineering 350, DOI 10.1007/978-3-319-19174-4_6

of CAD tools such as 3D placement and routing. In order to validate the architecture optimization models and performance of 3D Tree-based FPGA architecture, we used a two-tier 3D Tree-based interconnect model with seven Tree levels and arity four for each benchmarks circuits. The average improvement in speed measured for horizontally partitioned stacking methodology is 65.13 and 43.52 % for vertical partitioning method. The *horizontally* partitioned 3D stack methodology performed 1.7 times faster compared to 3D Mesh-based Industrial FPGA with identical logic resources [5]. The 3D Mesh-based FPGA reported in [5] with intermittent 2D and 3D switch blocks distribution estimated an average speed improvement of 38.3 % for identical logic density and array size.

6.2 Tree-Based FPGA Interconnect Architecture

The Tree-based FPGA architecture unifies two unidirectional networks using *Butterfly Fat-Tree* or (BFT) based network topology. Figure 6.1 shows the two-level Tree-based FPGA architecture with two unidirectional interconnect network. The downward network allows to connect switch blocks to LBs (leaves) inputs. The upward network uses a BFT based Tree to connect LBs outputs to Switch Blocks. The UMSBs in upward interconnect network allow LBs outputs to reach a larger number of Downward MSBs (DMSBs) and to reduce fanout on a feedback lines. The UMSBs are organized in a way that allows logic blocks (LBs) belonging to the same *owner cluster* (at level 1 or above) to reach exactly the same set of DMSBs at each level. Therefore, I/O pads, clusters or logic blocks positions inside the direct owner cluster become equivalent and we need no more to re-arrange them. For example in Fig. 6.1, an LB ouput can reach all 4 DMSBs of its owner cluster at level 1 and all the 16 DMSBs of its owner cluster at level two. As shown in Fig. 6.1, inputs and output pads are grouped together in specific clusters. The cluster size and the

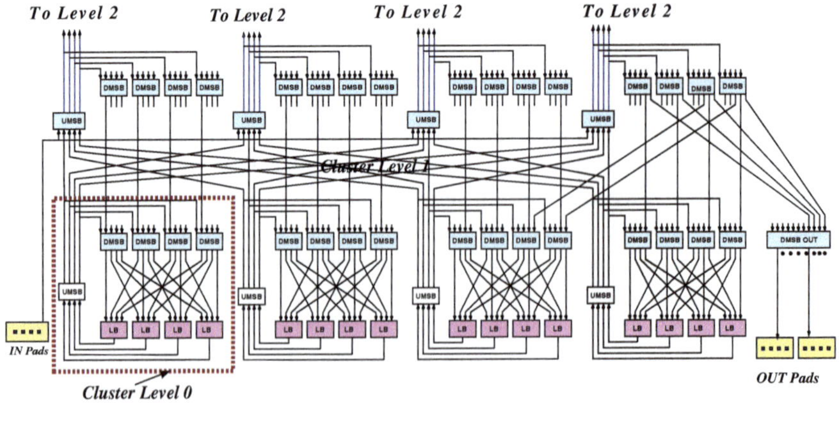

Fig. 6.1 Tree-based interconnect: upward and downward networks

level where it is located can be modified to obtain the best design fit. Each input pad is connected to all UMSBs of the upper level. In this way each input pad can reach all LBs of the architecture through different paths. Similarly, output pads are connected to all DMSBs of the upper level and this will allow it to reach all LBs through different paths. The I/O pads have flexible interconnection compared to LBs.

The Tree-based FPGA architecture with two BFT based unidirectional interconnection network may improve the flexibility and routability, but it has many issues in terms of physical implementation and performance. The Rent's growth rate ($p = 1$) is very penalizing in terms of wiring requirement. Interconnect wiring takes more area and dominates the array size. The wiring topology should be chosen to balance interconnect bandwidth with array size and expected design interconnect requirement. Thus we must control clusters signals bandwidth based on Rent's Rule [6]. Secondly the downward and upward interconnect switch network, and the LB output fanout depends on the number of levels. Thus in a large Tree-based FPGA architecture with high number of levels, performance may be penalized in term of larger wire and switch delays. As we know from 2D physical design experiments, the wire delay increases exponentially as the Fat-Tree grows to higher levels. To improve the interconnect area overhead and reduce wire delay, we proposed two different network partitioning methods, namely *Horizontal* and *Vertical* and 3D implementation discussed in Chap. 8 and also illustrated in Fig. 8.12. The main objective of network partitioning and 3D implementation is to optimize the network delay and routing resources of the 3D Tree-based FPGAs.

6.2.1 2D Tree-Based Interconnect: A Comparison with 2D Mesh-Based Interconnect

We have seen numerous studies [1, 2, 5, 7, 8] shows that the switch blocks (SBs) is the most area-consuming unit compared to other design elements in 2D Mesh-based FPGAs and this situation is becoming even worse in 3D Mesh-based FPGAs because the TSVs are located on 3D-SBs. Although the design and manufacturing engineers are trying to reduce TSV dimensions, the minimum feature size on the die is also shrinking. Therefore, the TSVs are expected to remain larger than wire dimensions in metal layers within the die [3, 9]. Moreover it has been reported in [4] that the TSV utilization is actually quite low if the 3D-SBs are with full vertical connectivity in use. The experimental and simulation results presented in recent publications point out that the utilization of TSVs is actually very low in 3D Mesh-based FPGAs [4] with full vertical connectivity due to sub-optimal Mesh-based FPGA designs, which motivates us to explore new interconnect topology and architecture with better optimization flexibility to achieve higher speed, low power consumption, reduced silicon footprint and increased logic density to minimize the performance gap between FPGAs and ASICs. Hence we selected Tree-based FPGA architecture with BFT based interconnect topology for the 3D design and implementation of high density FPGAs. Previous works [10–12], also confirms the multilevel BFT based 2D interconnect topology is able to reduce 59 % of the total number of switches and save 56 % of the total FPGA area compared to Mesh-based FPGA with identical logic density and array size [10]. Considering the challenges associated with 2D physical design of Tree-based FPGA [13, 14], we proposed two different network partitioning methodology to design and implement high density 3D FPGAs based on Tree-based interconnect network. The main focus of this chapter is on the discussion of the complete set of tools and technologies needed to conduct 3D design feasibility study and interconnect network characterization methodologies to build high performance 3D re-configurable systems based on Tree-based interconnect and a comparison procedure has been put in place and the end to validate the advantages of 3D Tree-based FPGA over 3D Mesh-based FPGA architectures.

The Tree-based FPGA architecture unifies two unidirectional interconnect network as illustrated in Fig. 6.1. The downward interconnection network is based on a butterfly-fat-tree style interconnect topology with a linearly populated downward mini switch boxes (DMSBs) and unidirectional wires. Similarly the upward interconnect network uses upward mini switch boxes (UMSBs) to connect logic block outputs to all DMSBs and further to higher levels of the Tree. The number of DMSBs of a cluster located at level ℓ is equal to the number of inputs of a cluster located at level $\ell - 1$. The upward network also uses BFT topology to connects LBs outputs to the DMSBs at each level. As shown in Fig. 6.1, we use UMSBs (Upward MSBs) to allow LBs outputs to reach a large number of DMSBs and to reduce fanout on feedback lines. The number of UMSBs of a cluster located at level ℓ is equal to the number of outputs of a cluster located at level $\ell - 1$. UMSBs are organized in a way allowing LBs belonging to the same *owner cluster* to reach exactly the same set of DMSBs at each level. Thus positions, inside the same cluster, are equivalent, and LBs can negotiate with their siblings the use of a larger number of DMSBs depending on their fanout.

As illustrated in Fig. 6.1, input and output pads are grouped into specific clusters and are connected to UMSBs and DMSBs, respectively. Thus, input pads can reach all LBs of the architecture, and output pads can also be reached by all the from different paths. Using UMSBs and DMSBs greatly enhances routability, but it increases the interconnect switches number. However this increase in number switches is compensated by reducing in/out signals bandwidth of clusters at every level using Rent's Rule [6] based interconnection optimization tool. In fact, netlists implemented on FPGA architecture often communicate locally (intra-clusters) and this fact can be exploited to reduce the bandwidth of signals with inter-clusters communication. A good estimation of netlists communication locality is given by Rent's Rule [6].

Based on this estimation authors in [15] showed that most netlist Rent's parameters range between 0.5 and 0.65. If most of the connections are routed locally and only a few of them communicate to the exterior of a local region, p will be expected to be small. In Tree-based architecture, both the upward and downward interconnects populations depend on this parameter. As shown in Fig. 6.2, we can depopulate the routing interconnect by reducing the number of inputs of each cluster of level 1

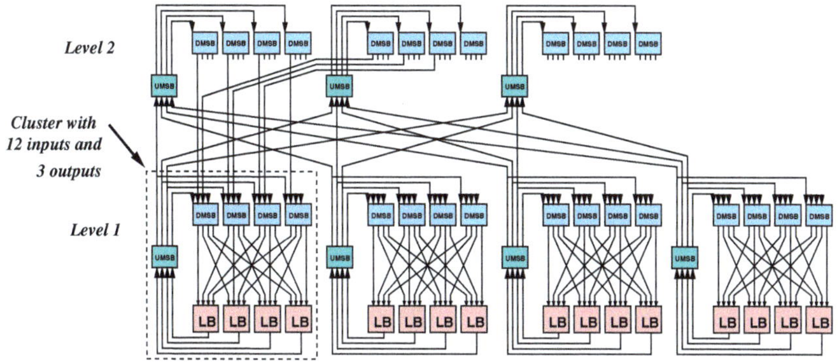

Fig. 6.2 An illustration of Tree-based interconnect optimization using Rent rule (Level ℓ with $p = 0.73$), based on Rent optimization, certain number inputs and outputs can be removed independently at each Tree levels

from 16 to 10 and outputs from 4 to 3 ($p = 0.734$). In this case, if we consider an architecture with 2 levels of hierarchy, we get a reduction in interconnect switches number from 521 to 368 (28 %). However, a reduction in the value of p reduces the routability of the architecture too. Thus we must find the best tradeoff between interconnect population and logic blocks occupancy. As shown in [16], the best way to improve circuit density is to balance logic blocks and interconnect utilization. In the case of Tree-based FPGA architecture, interconnect occupancy is controlled by p and logic occupancy factor is controlled by N, where N is the total number of LBs in the Tree.

6.3 Tree-Based Interconnect Partitioning

Tree-based FPGA is an interconnect dominated device. Network partitioning is used to reduce the wire length and vertical interconnects in a 3D design. In this section, two types of network partitioning methodologies are defined for the design and implementation 3D Tree-based FPGA as illustrated in Fig. 6.3: (1) *vertical partitioning:* the programmable interconnect network is partitioned vertically by placing the break-point at the highest level ℓ_v of the Tree-based programmable interconnect network to balance the silicon area and power consumption across multiple tiers of the 3D chip and (2) *horizontal partitioning:* the main objective is to optimize the critical path delay and improve logic density. The horizontal break-point is placed at a particular tree level ℓ_h based on the design and manufacturing constraints to achieve interconnect delay optimization using TSVs. The location of the level ℓ_v is always fixed at highest tree level, however the location of level ℓ_h is decided based on the architecture and wire delay requirements. Figure 8.12 illustrate both *vertical* and *horizontal break-points* in homogeneous Tree-based FPGA. As discussed already the main goal of horizontal partitioning is to limit the exponential

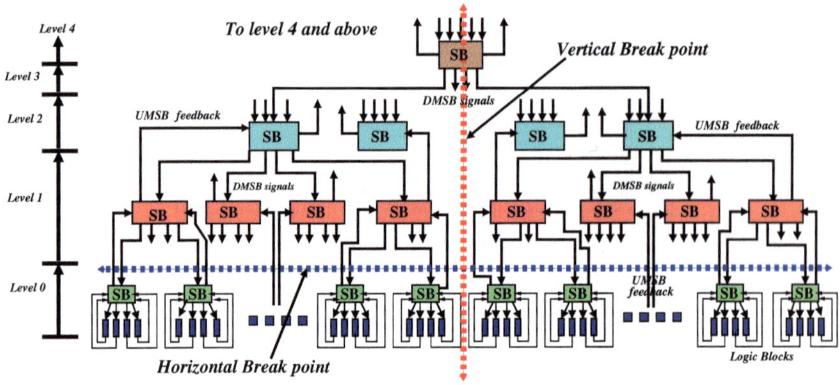

Fig. 6.3 Representation of programmable interconnect network partitioning (break point) of a four-level Tree-based homogeneous FPGA. *Horizontal* break-point: *blue dotted line*, *Vertical* break-point: *red dotted line*

increase in network delay as we increase the number of levels in the fat-Tree. In a horizontally-partitioned design the LBs and local programmable interconnects along with configuration memory is place in one layer and rest of the programmable interconnects and I/Os above the *break-point* along with configuration memory is placed in another layer. The advantage of such partitioning is that it provides flexibility in increasing logic density and optimizing network delay. However the horizontal partitioning methodology is not balanced in terms of area and power consumption. Previous work [11, 17] confirmed, that placing Hard-blocks like DSP slice and memory units at the higher level of Tree is more beneficial to optimize area and speed of heterogeneous FPGA. Figure 6.4 shows the heterogeneous version of Tree-based FPGA with the placement of hard-blocks and partitioning methods. In this case we utilized the white space left in active layer two to places hard-blocks to design heterogeneous Tree-based FPGA.

6.3.1 Vertical Partitioning

The main focus of vertical partitioning method is to balance the total silicon area and power consumption of the two-tier 3D homo/heterogeneous Tree-based FPGA equally between the active layers of the 3D stacked chip. The total number of LBs plus HBs and SBs are equally partitioned into multiple stacked tiers. The break-point is set at the highest level of the Tree and interconnected using TSVs as illustrated in Figs. 6.3 and 6.4. Since the break-point is set at highest Tree level, the total number of TSVs required for a vertically partitioned 3D test chip with seven-Tree levels and 16K LUTs is 40,960 for a fully connected (Rent = 1) two-tier Tree-based FPGA. However we used Rent's Rule [6] based interconnect optimization model to find optimum

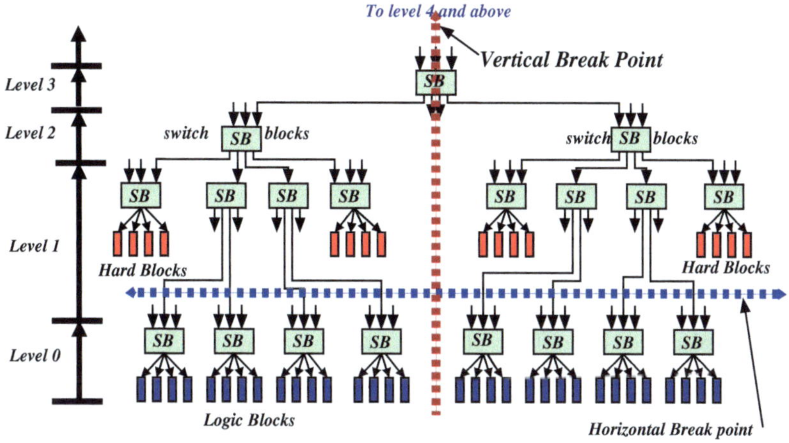

Fig. 6.4 Representation of programmable interconnect network partitioning (break point) of a four-level Tree-based heterogeneous FPGA. *Horizontal* break-point: *blue dotted line*, *Vertical* break-point: *red dotted line*, only downward network shown

number of TSVs and routing resources required for the design. The interconnect delay increases exponentially as the Tree grows to higher levels, the longest wire in 3D Tree-based FPGA located at highest Tree level is replaced by TSV and a limited wire length optimization is possible at other levels due the hierarchical nature of upward interconnection network. In a vertically partitioned Tree, the downward interconnection network is more localized inside the partitioned layout, however BFT-based upward hierarchical interconnect network connects feedback to all cluster inputs. In a vertically partitioned Tree-based interconnect, only 50 % of the upward interconnects are realized using TSVs and other 50 % uses local interconnects. This makes the *vertically partitioned* 3D FPGA $\approx$ 3.3 times slower in speed compared to *horizontally partitioned* 3D Tree-based FPGA. Nevertheless, the advantages of vertical partitioning method compared to horizontal includes reduced chip area by 50 %, balanced power consumption across the tiers, reduced number of TSVs, and design complexity is minimized.

6.3.2 Horizontal Partitioning

In *horizontal* partitioning methodology, the location of the *break-point* is decided based on the estimated interconnect network delay. The interconnect delay of Tree-based architecture increases exponentially [10, 14, 18] as the Tree grows to higher levels. Horizontal partitioning methodology is introduced optimize the exponential rise in Tree network delay as the Tree grows to higher levels. In horizontal partitioning methodology, the LBs and local interconnect levels below the *break-point* are placed in the bottom tier and programmable interconnect resources at levels above the break-point and I/O pads are placed in the top tier of the 3D stacked chip as illustrated in Figs. 6.3 and 6.4. The location of break point is decided based the delay measurements of tree levels starting from level 0 to the highest level of the tree interconnect. The delay measurements show the path delay of the programmable routing resources placed at the top tier of the partitioned layout reduces 3 times and an overall path delay reduces 3.3 times compared to 2D layout as discussed in Sect. 8.7 of Chap. 8. The interconnect path delay optimization is achieved by exploiting the design flexibility introduced by the segregation LBs and programmable routing resources into multiple tiers of the 3D design. Effectively almost 80 % of programmable routing network is placed on top of logic blocks is strongest point of *horizontal partitioning* methodology. The 3D test chip contains 16K LBs placed in tier 1 (bottom tier) with 65,536 vertical input pins and 16,384 vertical output (feedback) pins. For a fully connected 3D test chip requires 81,920 TSV communication between top (tier 0) and bottom tier (tier1).

6.3.3 Through Silicon via (TSV) Modeling

The 3D physical design section used six metal 130 nm technology node provided by Global Foundries (GF-130 nm) with the modified technology libraries to include TSVs according to the specification of $Tezzaron$ Semiconductor. The GF-130 nm/Tezzaron process has one polysiliocn layer and 6 metal layers, however the last metal layer is reserved for wafer-to-wafer or TSV connections. Therefore, only five metal layers are available for design and all of them are thin metal layers intended for digital routing. Tezzaron's via-first 3D manufacturing process produces very small TSVs that are approximately 1.2 μm wide with 2.5 μm minimum pitch and 6 μm height [3]. The liner thickness is 100 nm and we used SiO_2 for liner deposition. The estimated values provided by Tezzaron for TSV resistance R_{TSV} and capacitance C_{TSV} are ≈600 mΩ and 15*f*F respectively. The wire delay estimation of tree levels for the 3D stacked Tree-based FPGA is extracted from the two-tier layout developed using Tezzaron Process and validated using Mentor's spice accurate circuit simulator *Eldo*. The break-point TSV delay is measured using the TSV model presented in [2, 19] and illustrated in Fig. 6.5. The TSV delay estimated using *eldo* is ≈28–32 pS. The wire delay estimation of tree levels for the 3D stacked Tree-based FPGA is extracted from the two-tier layout developed using Tezzaron Process and validated using Mentor's spice accurate circuit simulator *Eldo*. The break-point TSV delay is validated using *eldo* is ≈28–32 pS for Tezzaron TSVs. In tier 0, the spatial distribution of TSVs, SBs and HBs are rearranged in order to optimize the wire delay and temperature distribution at higher levels.

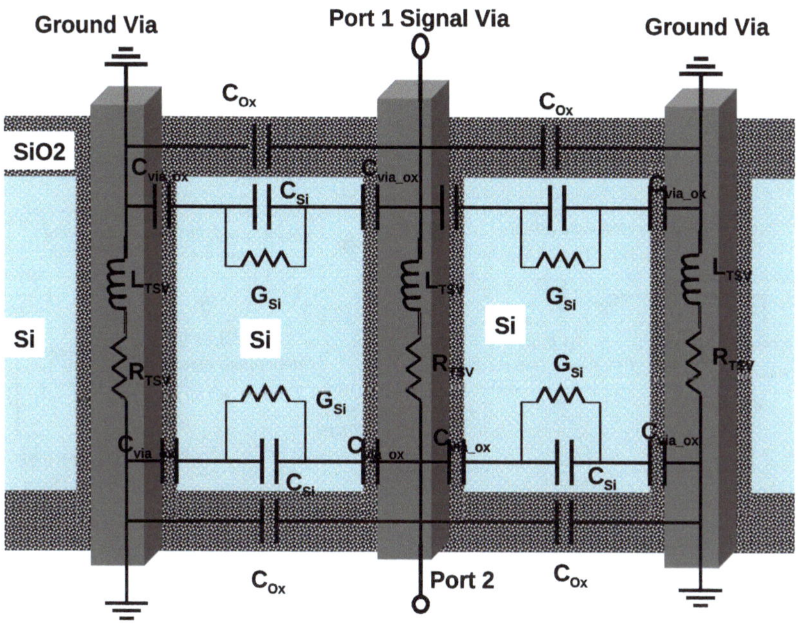

Fig. 6.5 Vertical interconnect (TSV) characterization model

6.4 3D Tree-Based Interconnect Optimization Methodology

In this section we discuss the interconnect optimization methodology and simulation results of a two-tier Tree-based FPGA architecture partitioned and stacked using *Horizontal* and *Vertical* network partitioning methodology. The main objective is to optimize TSV count and programmable routing resources in 3D Tree-based FPGA. Experiments are performed individually for each netlist using the optimization flow presented in Fig. 6.6. The architecture optimizer designed as an add-on utility of the main CAD flow presented in Chap. 5. The optimization program uses the router program implemented using the *pathfinder* algorithm [10, 18, 20], which uses an iterative, negotiation-based approach to successfully route all nets in an application netlist. The router program in association with a binary search algorithm, considers the same architecture with different p values at each levels of the two-tier 3D Tree-based FPGA to determine the smallest number of input and output signals at each Tree levels by allowing to route the benchmark circuits. At first, the optimization program considers architecture *break-point* level with different Rent (p) values. The purpose is to find, for all benchmark circuits, the architecture with the fewest necessary TSVs between the break-point levels while keeping the programmable interconnect resources placed in tier 0 and 1 intact. The solution provides the spatial distribution and minimum number of vertical interconnects required to route each benchmark in the two-tier Tree-based FPGA. From this solution we extract the

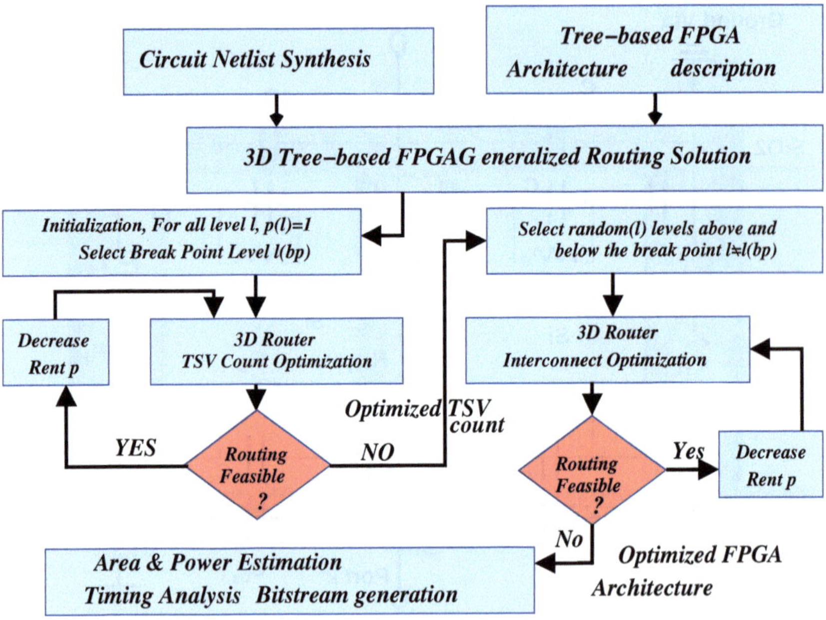

Fig. 6.6 TSV count and interconnect optimization flow

minimum possible number and location of TSVs that can removed from the architecture without compromising the performance of the 3D chip. The decision to remove TSVs is taken based on the spatial distribution and p values of all benchmark used in the optimization process. The highest p value obtained from all benchmarks at each levels will be set as the architecture Rent. To make 3D Tree-based FPGA more efficient in terms of design and manufacturing, it is essential to minimize the TSV count because TSV consumes more silicon area than horizontal interconnects [21]. After completing the break-point optimization, we use the Rent's parameter [6, 15] to optimize the programmable routing resources that are placed in tier 0 and 1 using random approach, in which the interconnect levels are selected randomly and modifly its inputs and outputs signals depending on the previous result obtained at the same level.

6.5 Interconnect Optimization: Homogeneous Tree

The Rent's parameter p defined for a Tree-based architecture is illustrated in Eq. 6.1. The Tree level is represented as ℓ and k is the cluster arity, c is the number of in/out pins of an LB and IO is the number of in/out pins of a cluster located at level ℓ. The optimization of upward and downward networks based on Rent's parameter is done as follows.

$$IO(\ell) = c.k^{\ell.p} \tag{6.1}$$

6.5.1 The Downward Programmable Network Model

As described in Fig. 3.13, the Tree-based FPGA architecture unifies two unidirectional upward and downward interconnection networks using a BFT based network topology to connect Downward MSBs (DMSBs) and Upward MSBs (UMSBs) to LBs inputs and outputs. A cluster stationed at level ℓ contains $N_{in}(\ell-1)$ DMSBs, where $N_{in}(\ell)$ is the number of inputs of cluster located at level ℓ with k outputs and $\frac{N_{in}(\ell)+kN_{out}(\ell-1)}{N_{in}(\ell-1)}$ inputs, whereas k is also the cluster arity size. Since DMSBs are full crossbar devices, the total number of switches at level ℓ cluster is $k(N_{in}(\ell)+kN_{out}(\ell-1))$. At each level ℓ, $\frac{N}{k^{\ell}}$ clusters, whereas N is total number Logic Blocks and the total number of switches in the downward network is

$$\sum_{\ell=1}^{\log_k(N)} k \times N \times \frac{N_{in}(\ell)+kN_{out}(\ell-1)}{k^{\ell}} \tag{6.2}$$

Following Eq. 6.1, we can simplify the number of outputs of a Logic Block is $N_{out}(0) = c_{out}$ and the number of inputs equal $N_{in}(\ell) = c_{in}.k^{\ell.p}$ and $N_{in}(\ell-1) = c_{out}.k^{(\ell-1)p}$ and so on. The total switches used at each level ℓ can be calculated by Eq. 6.3.

$$N_{interconnects}(down) = N \times (k^{p}c_{in} + kc_{out}) \times \sum_{\ell=1}^{\log_k(N)} k^{(p-1)(\ell-1)} \tag{6.3}$$

6.5.2 The Upward Programmable Network Model

Similar to the downward interconnect network. The upward interconnect network also built using a *Butterfly-Fat-Tree* network topology. In level ℓ every cluster contains $N_{out}(\ell-1)$ UMSBs with k inputs and outputs. UMBSs are also full crossbar devices with $k^2 \times N_{out}(\ell-1)$ switches at a level ℓ cluster. There are $\frac{N}{k^{\ell}}$ clusters at each level ℓ, and the total number of upward interconnection block is

$$\sum_{\ell=1}^{\log_k(N)} \frac{k^2 \times N}{k^{\ell}} \times N_{out}(\ell-1) \tag{6.4}$$

$N_{out}(0) = c_{out}$ is the outputs of Logic Block and using Eq. 6.1, $N_{out}(\ell - 1) = c_{out}.k^{(\ell-1)p}$. The total number of interconnect required for the upward interconnect network is calculated using Eq. 6.5

$$N_{interconnects}(up) = N \times k \times c_{out} \times \sum_{\ell=1}^{\log_k(N)} k^{(p-1)(\ell-1)} \tag{6.5}$$

The total number of Tree-based interconnect switches in Tree-based FPGA architecture is

$$N_{switch}(Tree) = N_{switch}(down) + N_{switch}(up)$$

$$N_{switch}(Tree) = N.(k^p c_{in} + 2kc_{out}) \sum_{\ell=1}^{\log_k(N)} k^{(p-1)(\ell-1)} \tag{6.6}$$

The number of switches per Logic Block is

$$N_{switch}(LB) = (k^p c_{in} + 2kc_{out}) \times \sum_{\ell=1}^{\log_k(N)} k^{(p-1)(\ell-1)} \tag{6.7}$$

$$N_{switch}(LB) = \begin{Bmatrix} (k^p c_{in} + 2kc_{out}) \times \frac{1-N^{p-1}}{1-K^{p-1}} & \text{if } p \neq 1 \\ (k^p c_{in} + 2kc_{out}) \times log_k(N) & \text{if } p = 1 \end{Bmatrix}$$

$$N_{switch}(LB)_{Tree} = \begin{Bmatrix} \mathrm{O}(1) & \text{if } p \neq 1 \\ \mathrm{O}(log_k(N)) & \text{if } p < 1 \end{Bmatrix} \tag{6.8}$$

To compare the total number of switches in Mesh-based FPGA architecture we used the analytical model presented in [16]. The switch growth rate for Mesh-based FPGA architecture is represented in Eq. 6.9.

$$N_{switch}(LB)_{Mesh} = O(N^{(p-0.5)} \tag{6.9}$$

Equations 6.8 and 6.9 shows that in Tree-based FPGA architecture, the switch requirement grow more slowly compared to Mesh-based architecture. These results are encouraging to improve the manufacturability of high density FPGAs, especially when p is less than 1. However the routability factor is different for Tree-based architecture compared to Mesh-based FPGA.

The total number of interconnects at different levels of the Tree is calculated by substituting $p = 1$ in the Eq. 6.6, where N is the total number of logic blocks, c_{in} and c_{out} are the number of inputs and outputs of logic blocks, k is the arity, and p and ℓ are the Rent's parameter and tree interconnect level. However in normal cases the value of p ranges from 0.3 to 0.8. At first, the optimization program considers architecture break point level with different Rent's parameter (p) values. The purpose is to find, for all benchmark circuits, the architecture with the fewest necessary TSVs between the *break-point* levels. As described in [10], in a Tree-based FPGA the reduction in number of interconnects at *level* ℓ impacts the number of interconnects at *level* $\ell + 1$, since the number of DMSBs/UMSBs at *level* $\ell + 1$ is equal to the number of inputs/outputs of a cluster at *level* ℓ. Using Eqs. 6.1 and 6.6, the Rent's parameter p value and optimized TSV count and interconnect requirements are calculated for each iteration to optimize the number of signals at the *break-point* levels. Once the break-point optimization is completed, the optimizer randomly chooses other tree levels above or below the *break-point* to optimize the available routing resources. The highest p value obtained from all benchmarks at each levels will be set as the architecture Rent to allow all benchmarks to rout without compromising critical path delay. In this way we define an optimized architecture with rent set for each levels as shown in Eq. 6.10 that can implement all sets of benchmark circuits used in this experimentation.

$$P_{arch}(\ell) = Max\left(\underbrace{P(\ell)}_{circuits}\right) \tag{6.10}$$

Table 6.1 presents the TSV count optimization results of two-tier Tree-based FPGA using horizontal partitioning method. A minimum possible reduction of 45 and 42 % TSVs observed for horizontal and vertical partitioning methodology. The reduction of TSVs and routing resources from other Tree levels can reduce routability and diminish the performance of the two-tier FPGA chip. The optimization tool accepts constraints regarding the critical path delay, since FPGA architecture optimization is trade-off between speed and area. The estimated speed degradation for horizontal and vertical partitioning methods are 4.44 and 3.2 % respectively from the fully connected FPGA architecture with Rent $p = 1$. A similar experiment with 3D Mesh-based FPGA architecture with identical logic density [2, 5] with 40 % reduction of TSV resulted in speed degradation of 11.5 % as illustrated in Table 6.1, which indicates the impact of TSV and routing resources optimization on speed is minimized in Tree-based FPGA compared to Mash-based FPGA.

Table 6.2 presents the results from TSV and architecture optimization experiments on each interconnect level of the Tree-based 3D FPGA. A minimum reduction of 45 and 42 % TSVs are recorded for horizontal and vertical break-point. An average speed degradation of 4.44 and 3.2 % is recorded in horizontal and vertical break-point. The optimized interconnect and TSV area for individual interconnect levels are reported in Table 6.2. Using our optimization flow, overall interconnect area of the 3D Tree-based FPGA is reduced by 40 %, which improves the manufacturability of 3D stacked Tree-based FPGA and also a cost effective solution to design and build high density 3D FPGAs. The test chip we designed using 3D design process has seven Tree levels with arity set to 4. It includes 16K LUTs and the horizontal break-point placed between levels 3 and 4 of the BFT based horizontally partitioned interconnect network. We have 65,536 3D nets named as cluster inputs pins and 16,384 feedback networks

Table 6.1 Two-tier Tree-based FPGA: TSV count optimization results

	Tree levels = 7, Arity = 4, Arch = 4 × 4 × 4 × 4 × 4 × 4 × 4			
	Tree-based FPGA		Tree	Mesh
Circuits MCNC	Optimized Rent's "p"	$3D_{TSV}$ gain (%)	35 % TSV reduction speed degradation (%)	40 % TSV reduction speed degrade (%)
alu4	0.47	53	4.3	2.34
apex2	0.51	49	5.8	11
apex4	0.54	46	1.1	10
bigkey	0.49	51	2.8	4.1
clma	0.55	45	4.8	25
des	0.48	52	4.1	8
diffeq	0.48	52	4.5	−14
dsip	0.54	46	4.1	4
elliptic	0.52	48	3.4	34
ex1010	0.55	45	3.5	5
ex5p	0.53	47	5.1	12
frisc	0.52	48	5.4	28
misex3	0.54	46	5.2	−8
pdc	0.54	46	3.8	10
s298	0.53	47	5.8	19
s38417	0.53	47	5.1	8
s38584	0.52	48	4.5	9
seq	0.51	49	5.5	8
spla	0.51	49	5.2	6
tseng	0.53	47	4.8	7
ava	0.53	47	5.3	8.8
Max	0.55	53	5.8	28
Average			4.44	11.5
	Maximum interconnect requirement, p = 0.55			
	Minimum possible TSV reduction = 45 %			

pins. For a fully connected (Rent = 1) horizontally partitioned 3D homogeneous FPGA expected to have 81,920 3D nets required TSV communication excluding I/O pads. The vertically partitioned 3D test chip with 7 Tree levels and 16K LUTs require 40,960 TSVs for a fully connected (Rent = 1) two-tier test chip. The seven levels Rent = 1 homogeneous Tree-based FPGA architecture is defined using Eq. 6.11.

$$P_{arch}(\ell)_1 = \ell(0)_1 \times \ell(1)_1 \times \ell(2)_1 \times \ell(3)_1 \\ (break - point)\ell(4)_1 \times \ell(5)_1 \times \ell(6)_1 \tag{6.11}$$

Table 6.2 Architecture optimization results

Tree levels = 7	Arity = 4, Arch = $4 \times 4 \times 4 \times 4 \times 4 \times 4 \times 4$		
Tree-based architecture levels	3D chip active layer	Optimized Rent '*p*'	Optimized area (μm^2)
Logic blocks	Layer 1	–	93,635,273
Switch level 0	Layer 1	0.65	2412
Switch level 1	Layer 1	0.54	10,800
Switch level 2	Layer 1	0.66	37,496
$Breakpoint_{Horizontal}$ Switch level 3	Horizontal break point level 3, $p_H = 0.55$ $p_V = 0.58$ Area = 232,128		
Level 3–4	TSV area = 540,672 μm^2		
Switch-level 4	Layer 2	0.63	6,072,770
Switch level 5	Layer 2	0.64	45,553,499
$BreakPoint_{Vertical}$ Switch level 6	Vertical break point level 6 $p_V = 0.58, p_H = 0.65$ Area = 42,139,683		
Level 6	TSV area = 28,508 μm^2		
Speed degradation	Vertical = 3.2 %, horizontal = 4.4 %		

The optimized Tree-based FPGA architecture for horizontal partitioning methodology is defined using Eq. 6.12.

$$P_{arch}(\ell_H)_{optimal} = \ell(0)_{0.65} \times \ell(1)_{0.54} \times \ell(2)_{0.66} \times \ell(3)_{0.55} \\ (b-p)\ell(4)_{0.63} \times \ell(5)_{0.64} \times \ell(6)_{0.65} \tag{6.12}$$

And the optimized Tree-based FPGA architecture for vertical partitioning methodology is defined using Eq. 6.13.

$$P_{arch}(\ell_V)_{optimal} = \ell(0)_{0.65} \times \ell(1)_{0.54} \times \ell(2)_{0.66} \times \ell(3)_{0.58} \\ (b-p)\ell(4)_{0.63} \times \ell(5)_{0.64} \times \ell(6)_{0.58} \tag{6.13}$$

The Eq. 6.13 shows the optimized Rent set for each levels in the FPGA architecture. With help of TSV and interconnect optimization, we removed 17,203 TSVs from vertically and 36,864 TSVs from horizontally partitioned two-tier 3D test chips with minimal impact on speed. We compared our simulation results with multi-stack 3D Mesh-based FPGA using the methodology reported in [2, 5] and observed a speed degradation of 11.5 for 40 % reduction in TSV count with identical array size and logical resources.

6.6 Heterogeneous Tree-Based FPGA Architecture

The architecture of 3D heterogeneous Tree-based FPGA is significantly more complex with multilevel BFT based routing resources and LBs with additional features such as adders and multipliers. Since the benchmark circuits plays a major role in the heterogeneous FPGA exploration, three sets of benchmarks [11, 22–25] were chosen based on trends of communication between different types of hard-blocks. Generally in academia and industry, the quality of an FPGA architecture is measured by mapping a certain set of benchmarks on it. Thus the selection of benchmarks plays a very important role in the exploration of heterogeneous FPGAs. This work puts special emphasis on the selection of benchmark circuits, as different circuits can give different results for different architecture techniques. This work categorizes the benchmark circuits by the trend of communication between different blocks of the benchmark. So, three sets of benchmarks are assembled having distinct trend of inter-block communication. These benchmarks are shown in Tables 6.3, 6.4 and 6.5. Benchmarks shown in Table 6.3 are developed at [22], the benchmarks shown in Table 6.4 are obtained from [24] and the benchmarks shown in Table 6.5 are obtained from [23]. The communication between different blocks of a benchmark can be mainly divided into the following four categories:

1. CLB-CLB: CLBs communicate with CLBs.
2. CLB-HB: CLBs communicate with HBs and vice versa.
3. HB-HB: HBs communicate with HBs.
4. IO-LB/HB: I/O blocks communicate with CLBs and HBs.

In SET I benchmarks, the major percentage of total communication is between HBs (i.e. HB-HB) and only a small part of total communication is covered by the communication CLB-CLB or CLB-HB. On average, in SET I, the HB-HB communication

Table 6.3 DSP Benchmarks SET I

Circuit name	Inputs	Outputs	LUTs (LUT-4)	Mult (8×8)	Slansky ($16 + 16$)	Sff (8)	Sub ($8 - 8$)	Smux (32:16)	Function
ADAC	18	16	47	–	–	2	–	1	–
DCU	35	16	34	1	1	4	2	2	Discrete cosine transform
FIR	9	16	32	4	3	4	–	–	Finite impulse response
FFT	48	64	94	4	3	–	6	–	Fast fourier transform

Table 6.4 Open Core Benchmarks Set II

Circuit name	No of Inputs	No of outputs	No of LUTs (LUT-4)	No of multipliers (16 × 16)	No of adders (20 + 20)	Function
cf_fir_3_8_8_open	42	18	159	4	3	Finite impulse response (8 bit)
cf_fir_7_16_16	146	35	638	8	14	Finite impulse response (16 bit)
cfft16x8	20	40	1511	–	26	–
cordic_p2r	18	32	803	–	43	Polar to rectangular
cordic_r2p	34	40	1328	–	52	Rectangular to polar
fm	9	12	1308	1	19	–
fm_receiver	10	12	910	1	20	–
lms	18	16	940	10	11	–
reed_solomon	138	128	537	16	16	–

Table 6.5 Open Core Benchmarks Set III

Circuit name	No of inputs	No of outputs	No of LUTs (LUT-4)	No of multipliers (18 × 18)	Function
cf_fir_3_8_8_ut	42	22	214	4	Finite impulse response (8 bit)
diffeq_f_systemC	66	99	1532	4	–
diffeq_paj_convert	12	101	738	5	–
fir_scu	10	27	1366	17	–
iir1	33	30	632	5	Infinite impulse response (16 bit)
iir	28	15	392	5	Infinite impulse response (8 bit)
rs_decoder_1	13	20	1553	13	Decoder
rs_decoder_2	21	20	2960	9	Decoder

takes up to 80 % of the total communication between different instances of the benchmarks (netlists). Similarly, in SET II the major percentage of total communication is HB-CLB and in SET III, major percentage of total communication is covered by CLB-CLB. Normally the percentage of IO-CLB/HB is a very small part of the total communication for all the three sets of benchmarks.

6.6.1 Interconnect Optimization: Heterogeneous Tree

The TSV and architecture optimization are performed based on Rent's parameter [6, 15] p defined for a Tree-based heterogeneous architecture as shown in Eq. 6.14. The Tree level is represented as ℓ and k is the cluster arity, c is the number of in/out pins of an LBs, a_x is the number of in/out pins of a HBs of type x, ℓ_x is the level at which the HBs are located, b_x is the number of HBs at that level, z is the number of HBs supported by the architecture and IO is the number of in/out pins of a cluster located at level ℓ. Since there are more than one type of HBs, their contribution is accumulated and then added to the *LB(p)* of Eq. 6.14 to calculate p. The value of p determines the total number of interconnects at each level of the Tree-based architecture and it is averaged across all the levels to determine the p for the architecture.

$$IO = \left(\underbrace{c.k^{\ell}}_{LB(p)} + \underbrace{\sum_{x=1}^{z} a_x.b_x.k^{(\ell-\ell_x)}}_{HB(p)} \right)^{p} \tag{6.14}$$

$$HB(p) = \begin{cases} 0 & \text{if } (\ell - \ell_x < 0) \\ a_x.b_x.k^{(\ell-\ell_x)} & \text{if} (\ell - \ell_x \geq 0) \end{cases} \tag{6.15}$$

As described in in Chap. 3 [12, 18], in a Tree-based FPGA the reduction in number of inputs/outputs of the clusters at *level* ℓ impacts the number of routing resources at *level* $\ell+1$, since the number of DMSBs/UMSBs at *level* $\ell+1$ is equal to the number of inputs/outputs at *level* ℓ. At first, the optimization program considers architecture *break-point* level with different *Rent* (p) values. The purpose is to find, for all benchmark circuits, the architecture with the fewest necessary TSVs between the break-point levels while keeping the programmable interconnect resources placed in tier 0 and 1 intact. The solution provides the spatial distribution and minimum number of vertical interconnects required to route each benchmark in the two-tier heterogeneous Tree-based FPGA. From this solution we extract the minimum possible number and location of TSVs that can be removed from the two-tier architecture without compromising the performance of the 3D chip in terms of speed and routability. The decision to remove TSVs is taken based on the spatial distribution and p values of all benchmark used in the optimization process.

The Table 6.6 presents the highest rent set for each level in our two-tier Tree-based heterogeneous FPGA architecture. The test chip we designed using 3D design process has seven Tree levels with arity set to 4. It includes 16K LUTs and the horizontal *break-point* placed between Tree levels 3 and 4 of the BFT based horizontally partitioned interconnect network. We have 65,536 3D nets named as cluster inputs pins and 16,384 feedback networks pins. For a fully connected (Rent = 1) horizontally partitioned 3D heterogeneous FPGA expected to have 81,920 3D nets required TSV communication excluding I/O pads. The vertically partitioned 3D test

Table 6.6 Architecture optimization results

Tree levels = 7	Arity = 4, Arch = 4 × 4 × 4 × 4 × 4 × 4 × 4		
Architecture levels	3D chip layer	Optimized Rent 'p'	Optimized area (μm^2)
Logic blocks	Layer 1	–	93,635,273
Switch level 0	Layer 1	0.67	2412
Switch level 1	Layer 1	0.58	10,800
Switch level 2	Layer 1	0.68	37,496
$Breakpoint_{Hori}$ Switch level 3	Horizontal break point Level 3 $p_{Horizontal} = 0.59, p_{Vertical} = 0.61$ Area = 232,128		
Level 3–4	TSV area = 4423.68 μm^2		
Switch level 4 + HBs	Layer 2	0.67	6,072,770
Switch level 5 + HBs	Layer 2	0.66	45,553,499
$Breakpoint_{Verti}$ Switch level 6 + HBs	Vertical break point Level 6 $p_{Horizontal} = 0.65, p_{Vertical} = 0.57$ Area = 42,139,683		
Level 6	TSV area = 19169.28 μm^2		

chip with 7 Tree levels and 16K LUTs require 40,960 TSVs for a fully connected (Rent = 1) two-tier test chip. Table 6.6 presents the TSV and interconnect optimization results. The minimum possible reduction of TSVs recorded for horizontal and vertical break-points are 41 and 43 % respectively. The seven levels Rent = 1 heterogeneous Tree-based FPGA architecture is defined using Eq. 6.16.

$$P_{arch}(\ell)_1 = \ell(0)_1 \times \ell(1)_1 \times \ell(2)_1 \times \ell(3)_1 \\ (break - point)\ell(4)_1 \times \ell(5)_1 \times \ell(6)_1 \tag{6.16}$$

The optimized FPGA architecture for horizontal partitioning is defined using Eq. 6.17.

$$P_{arch}(\ell_H)_{optimal} = \ell(0)_{0.67} \times \ell(1)_{0.58} \times \ell(2)_{0.68} \times \ell(3)_{0.59} \\ (b - p)\ell(4)_{0.67} \times \ell(5)_{0.66} \times \ell(6)_{0.65} \tag{6.17}$$

The optimized FPGA architecture for vertical partitioning is defined using Eq. 6.18.

$$P_{arch}(\ell_V)_{optimal} = \ell(0)_{0.67} \times \ell(1)_{0.58} \times \ell(2)_{0.68} \times \ell(3)_{0.61} \\ (b - p)\ell(4)_{0.67} \times \ell(5)_{0.66} \times \ell(6)_{0.57} \tag{6.18}$$

The Eqs. 6.17 and 6.18 presents the optimized Rent set for each levels in the heterogeneous Tree-based FPGA architecture. An average speed degradation using all four set of benchmarks are 1.7 and 0.8 % respectively for horizontal and vertical break-points. With help of TSV and interconnect optimization, we removed 17,612

TSVs from vertically and 33,587 TSVs from horizontally partitioned two-tier 3D test chips with minimal impact on speed. Using our optimization flow, total interconnect area of 3D Tree-based heterogeneous FPGA reduced by 35 % compared to the 2D Tree-based FPGA architecture.

6.7 Critical Path Delay Analysis

The critical path delay evaluation of vertical and horizontal partitioning methodology of 3D homogeneous and heterogeneous Tree-based FPGA architecture is performed using the experimental flow illustrated in Fig. 5.15.

6.7.1 Delay Analysis: Homogeneous Tree

To evaluate the performance of the proposed 3D Tree-based FPGA architecture, we place and route the largest set of MCNC benchmark circuits [26], and compare this with the 3D Mesh-based FPGA architecture [2, 5]. We used the optimal two-tier Tree-based FPGA architecture partitioned using horizontal and vertical partitioning methodology illustrated in Eqs. 6.12 and 6.13 to analyze the reduction in critical path delay. The two-tier Tree-based FPGA architecture has 7 (0–6) levels. The horizontal *break-point* is placed between level 3 and 4 and vertical *break-point* is placed at the last level 6. To evaluate the critical path delay, we run the place and route program for each benchmark circuits. The delay analysis of vertical and horizontal break point two-tier Tree-based FPGA is presented in Table 6.7. The respective average reduction in delay measured for horizontally and vertically partitioned stacking methodology are 65.13 and 43.52 %. The horizontally partitioned 3D stacking methodology provides 1.5 times speed improvement compared to vertical partitioning method. The speed improvement in horizontal partitioning method is due to design optimization and minimization of interconnect wire length at the higher levels tree networks that are placed in tier 0 of the 3D stacked two-tier FPGA chip as described in Chap. 8. In tier 0 we have additional design flexibility to re-order programmable routing resources to optimize wire length. However in the vertical break-point method, the interconnects at the highest tree level is optimized using TSVs and the rest of tree levels only limited optimization possible due to hierarchical nature of upward and downward-interconnect network. The reduction critical path delay for 3D Tree-based FPGA compared to Mesh-based FPGA is presented in Table 6.7. The multi-layer 3D Tree-based FPGA interconnect using TSVs shows an average of 65.13 % speed improvement compared to the 2D counterpart. The 3D Mesh-based FPGA reported in [2, 5] with heterogeneous interconnect fabric using intermittent 2D and 3D switch blocks distribution with the same layout area measured an average speed improvement of 38.3 %. To conclude the comparison results presented in Table 6.7, the horizontally partitioned 3D Tree-based FPGA is 1.5 times

Table 6.7 Two-tier 3D Tree-based FPGA critical path delay analysis

P_{arch} =	$\ell(0)_{0.65} \times \ell(1)_{0.54} \times \ell(2)_{0.66} \times \ell(3)_{0.55}(BP)\ell(4)_{0.63} \times \ell(5)_{0.64} \times \ell(6)_{0.65}$					
	Critical path performance (nS)			Performance gain (%)		
Name MCNC	Tree-based 2D (nS)	Vertical 3D (nS)	Horizontal 3D (nS)	2D Versus 3D verti (%)	2D Versus 3D hori (%)	2D Versus 3D mesh (%)
alu4	59.91	41.73	25.81	30.33	56.91	44.23
apex2	80.41	45.18	30.92	43.81	65.54	55.41
apex4	76.42	46.61	31.83	38.99	58.34	51.74
bigkey	79.1	27.60	20.19	65.11	74.48	50.97
clma	198.6	90.33	59.48	54.38	69.96	50.98
des	90.8	40.36	28.83	55.55	68.25	38.07
diffeq	62.6	48.46	26.66	22.59	57.41	5.25
dsip	61.9	28.55	19.78	53.88	68.05	46.05
elliptic	107.1	83.73	42.76	21.75	60.02	28.49
ex1010	143.1	74.85	45.42	47.69	68.26	25.32
ex5p	168.2	64.71	41.43	61.53	75.37	61.75
frisc	129.6	82.28	42.82	36.51	66.96	17.11
misex3	67.4	41.38	24.94	38.61	63.00	48.95
pdc	143.9	69.04	45.86	52.02	68.13	55.71
s298	130.81	81.54	45.81	37.67	64.98	61.23
s38417	75.46	43.38	30.69	42.78	59.33	21.64
s38584	118	69.54	40.51	41.07	65.67	52.35
seq	64.58	42.91	24.59	33.56	61.92	44.00
spla	109.54	58.57	38.29	46.26	65.04	57.03
tseng	131.1	70.47	45.51	46.25	65.07	27.78
ava	211.5	161.63	111.21	23.58	47.42	0.0
Average	104.88	57.37	35.47	43.52	65.13	38.30

faster than 3D Mesh-based FPGA with identical array and logic density. The design and manufacturing solution presented in [5] by using same silicon area for both 2D and 3D-SBs is not piratical for high density FPGAs. This design style will increase silicon footprint of high density FPGAs, but the 3D multi-tier Tree-based FPGA with horizontal or vertical partitioning is more efficient as well as economical design and manufacturing methodology because in our 3D design we use uniform style switch blocks, which further reduces design complexity.

6.7.2 Delay Analysis: Heterogeneous Tree

In order to have a detailed path delay analysis and architecture optimization, we used the optimal vertical and horizontal Tree-based FPGA architecture obtained from architecture optimization methodologies presented in Eqs. 6.17 and 6.18. The *break-points* for horizontal partitioning method ℓ_h is set between Tree level 3 and 4, while for vertical partitioning, the *break-point* ℓ_v is set at the highest level of the Tree-based interconnect network. Tables 6.8, 6.9 and 6.10 presents the estimated critical path delay for 2D and 3D heterogeneous Tree-based FPGA. In all experiments, the cluster size and LB size set to 4. For horizontal partitioning method, the speed gain measured for SET I, II and III benchmarks are 51.7, 55.8 and 50.2 % respectively. There is a slight advantage for SET II benchmark, because the *horizontally* partitioned architecture is optimized for HBs to LBs communication. Nonetheless all 3 SETs

Table 6.8 Set I, Benchmark, heterogeneous FPGA experimental results

Digital signal processing (DSP) Benchmark Set I

Break point		Vertical partitioning	Horizontal partitioning
Circuit name	Hard blocks add/multi	2D versus 3D gain (%)	2D versus 3D gain (%)
ADAC	–/1	11.2	56.1
DCU	1/2	9.8	54.7
FIR	4/–	9.3	48.3
FFT	4/6	8.5	47.6
Maximum	6	11.2	56.1
Average	–	9.7	51.7

Table 6.9 Set II Benchmark, heterogeneous FPGA experimental results

Break point		Vertical partitioning	Horizontal partitioning
Circuit name	Hard blocks add/multi	2D versus 3D gain (%)	2D versus 3D gain (%)
	Open Core Benchmark Set II		
cf_fir_3_8_8	4/3	7.8	31.3
cf_fir_7_16_16	8/14	14.2	54.6
cfft16x8	–/26	18.9	65.4
cordic_p2r	–/43	13.8	54.5
cordic_r2p	–/52	21.3	66.1
FM	1/19	17.8	69.3
FM receiver	1/20	17.6	65.3
LMS	10/11	18.6	35.1
Reed_Solomon	16/16	16.8	61.1
Maximum	16/52	21.3	69.3
Average	–	16.3	55.8

Table 6.10 Set III Benchmark, heterogeneous FPGA experimental results

Break point		Vertical partitioning	Horizontal partitioning
Circuit name	Hard blocks add/multi	2D Versus 3D gain (%)	2D Versus 3D gain (%)
	Open Core Toronto VTR Benchmark Set III		
cf_fir_24_16_16	8	9.6	45.8
cf_fir_3_8_8	4	8.2	38.2
diffeq_f_systemC	4	19.7	67.5
diffeq_paj_convert	5	6.4	29.9
fir_scu	17	21.3	67.9
IIR1	5	9.8	35.5
IIR	5	11.3	36.2
RS_Decoder_1	13	18.6	65.7
RS_Decoder_1	9	23.6	72.8
oc54_cpu	27	13.3	44.8
sv_chip1_hierarchy_no_mem	152	12.8	48.6
stereovision0	66	15.5	66.8
stereovision1	50	14.8	64.5
stereovision2	231	19.3	58.5
Maximum	231	23.6	72.8
Average		14.1	50.2

performs well due the delay optimization performed in tier 0 and 1 of the 3D stacked chip. A Rent's Rule based statistical model for 3D heterogeneous Mesh-based FPGA presented in [27] shows 58 % performance improvement. However the paper lacks the hardware infrastructure to experiment and validate the achievement presented in it. The improvement in speed for horizontal partitioning arise from the design flexibility that has been introduced in tier 0, in which the programmable routing resources and HBs are placed. As explained in [4, 9, 27] the programmable interconnect overhead is main design element to increase the FPGA system performance. The main advantage of horizontally partitioned 3D heterogeneous Tree-based FPGA, is that, it offers design flexibility to optimize the wire length of higher levels of the Tree interconnects. For vertical partitioning method, the respective speed gains measured for SET I, II and III benchmarks are 9.7, 16.3 and 14.1 %. As described in Sect. 6.3.1, the vertically partitioned 3D heterogeneous FPGA is optimized to balance silicon area and power consumption requirements of the 3D design. However the delay reduction in *vertical* partitioning method is due the optimization performed on feedback networks.

6.8 LUT and Cluster Size Effect on Performance

The Tree-based FPGA is more area efficient in comparison to Mesh-based FPGA. The impact of cluster size and LUT (Look-Up-Table) size on area is been presented in [10, 18]. The experimental data analysis shows Tree-based FPGA architecture with cluster arity equals 4 and LUT size equal 4 is the best in terms area and speed [10, 18]. Figure 6.7 shows the how the switch block size is affected when we increase the cluster size of a Tree-based FPGA architecture. The 2D comparison between Tree-based FPGA and Mesh-based shows, around 59 % reduction in total number of switch requirement for Tree-based FPGA architecture, when we use cluster arity equals 4. In this section we evaluate the impact of LUT and cluster size on critical path delay of two-tier 3D Tree-based FPGA. A cluster is group of basic logic blocks (LBs) that are fully connected using a multiplexer-based crossbar as illustrated in Fig. 6.7 [28]. In Fig. 3.13 a cluster has four LBs and stated as arity-4, however in Fig. 6.7 shows the difference in size of SBs when we increase the cluster size from 4 to 8. Several studies in the past have examined the impact of LUT and cluster size on area and performance of 2D Mesh-based FPGAs. The work presented in [29] shows the LUT size of 4–6 and cluster size of between 3 and 10 provides the best area delay product for 2D Mesh-based FPGAs. Increasing LUT size or cluster size generally increase the functionality of the Logic blocks, which has many advantages: it decreases the total number of LBs needed to implement a given function and also decreases the number of such blocks in the critical path, whereby improving the FPGA system performance. The impact of LUT and cluster size on 2D Tree-based FPGA presented in [10]. The experimental results presented in [10] concludes that the LUT and cluster size of 4 provides the best area-delay product for a 2D Tree-based FPGA architecture. To evaluate the impact of LUT and cluster size on critical path delay of 3D two-tier Tree-based FPGA architecture, we fixed the cluster size to 4 while analyzing the impact of speed on LUT size variation and vice versa for cluster size experiment.

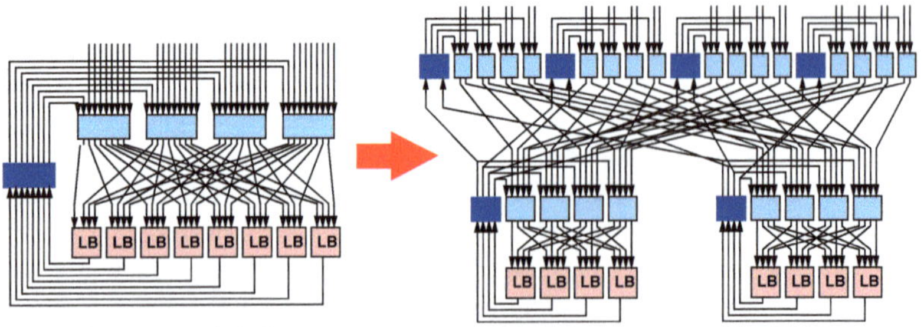

Fig. 6.7 Tree-based FPGA architecture with different cluster size

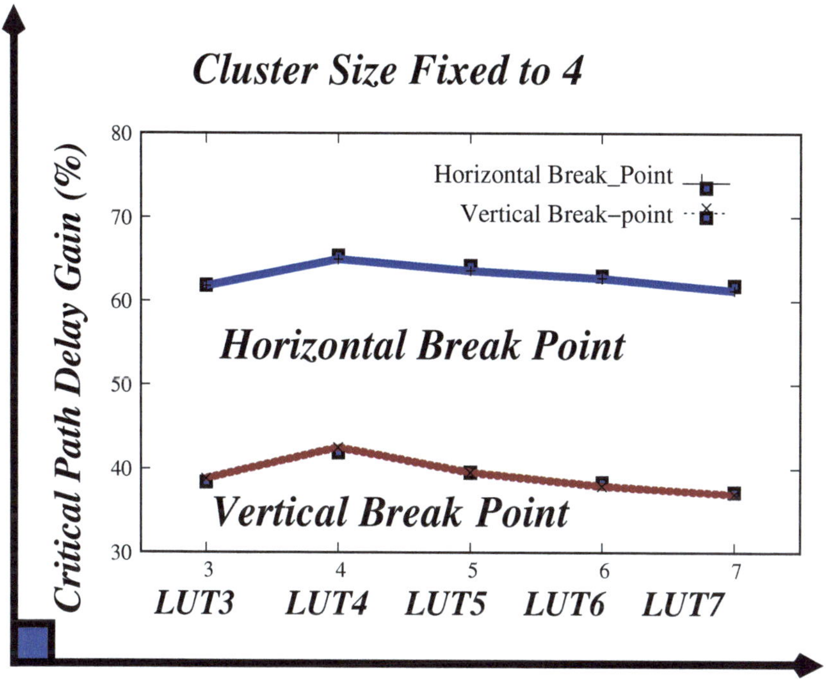

Fig. 6.8 Effect of LUT size on performance with cluster size fixed to 4

Figure 6.8 presents the impact of increasing LUT size from 3 to 7 with cluster size fixed to 4 on critical path delay of 3D Tree-based FPGA architecture using horizontal and vertical network partitioning methodology. As the LUT size increases, the area of chip and switch delay increases. The critical path delay analysis experiments consider the impact of increased switch delay, number of interconnects and TSVs as LUT size increases. The results shows that, LUT size = 4 has the best area-delay product as illustrated in Fig. 6.8. The conclusion we derive from these experiments is that, an increase in LUT size has an exponential effect on size of LBs [29]. As LUT size increase, the switch box size and interconnects requirement increases and also number TSV increases. This causes the LB and switch delay to increase and thus total delay and area increases. We also observed as the LUT size increase, the number of components in critical path reduces due to localization of routing resources, the delay of associated with LBs and crossbars increase and the speed improvement is diminished.

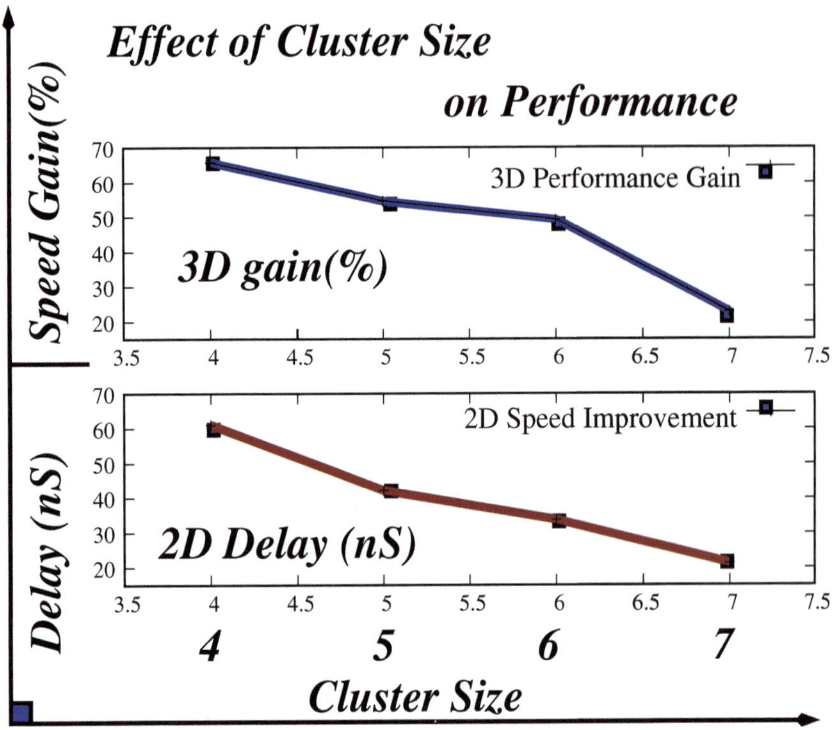

Fig. 6.9 Impact of cluster size on performance with LUT size fixed to 4

Figure 6.9 presents the effect of increasing cluster size from 4 to 7 with LUT size fixed to 4. As cluster size increases the buffers must be sized larger to withstand the larger loading from internal muxes, which result in an increase in the basic LB delay and also the size of each logic tile increases and therefore, the length of wires being driven by each buffer increases and this process increases the capacitive loading of each wire. To compensate this, we need to use larger channel width transistors in our buffer design. As cluster size increase the logic density increase and this forces the mapped application to use more local routing resources in the tree levels close to logic blocks than routing resources at higher tree levels in a timing driven routing procedure. One other fact is we use high cost factor for $level = 3$ TSV interconnects. This makes the critical delay shorter as cluster size increases. By varying the break-point location, the critical path delay of 3D Tree-based FPGA can be optimized for the horizontal partitioning method, however this process makes the architecture more application-specific. Our area and critical path delay analysis against various LUT and cluster size analysis reveals cluster and LUT size equal to 4 is better in terms of speed, power and silicon area to design and manufacture a genera-purpose high density and high speed 3D Tree-based FPGA systems.

6.9 Power Optimization

Power consumption is a major concern of designing CMOS integrated circuits in deep sub-micron technologies. As we move into deep-sub-micron geometries, increases in functionality per square millimeter come at the cost of higher static power consumption due to higher transistor leakage. This is especially the case of 3D FPGAs, because they are significantly less power efficient than custom integrated circuits (ASICs). Recent studies have shown that FPGAs are around 3 times less power efficient than ASICs [17, 30]. This power inefficiency is caused by the large programming overhead of FPGAs, including the pass-transistor switches, muxes, buffers, and configuration memory used in the programmable routing resource, which occupies 80 % of the silicon area [30–32]. Especially in the case of Tree-based FPGA, the wire length increases exponentially as the Tree grows to higher levels and the causes the programming overhead power to increase. From this discussion, it is clear that reducing wasteful interconnect power consumption in a Tree-based FPGA must involve reducing routed net lengths and/or programming overhead. These power reduction techniques cover BFT based interconnect network partitioning methodologies (*horizontal or vertical*), programmable interconnect layout design and optimization methods.

The power optimization of two-tier 3D stacked Tree-based FPGA is achieved through the minimization of TSV count and programmable routing resources. The optimized routing resources and TSV count presented in Fig. 6.10. In Mesh-based industrial 3D FPGA, the same power is used for individual blocks in multiple tiers

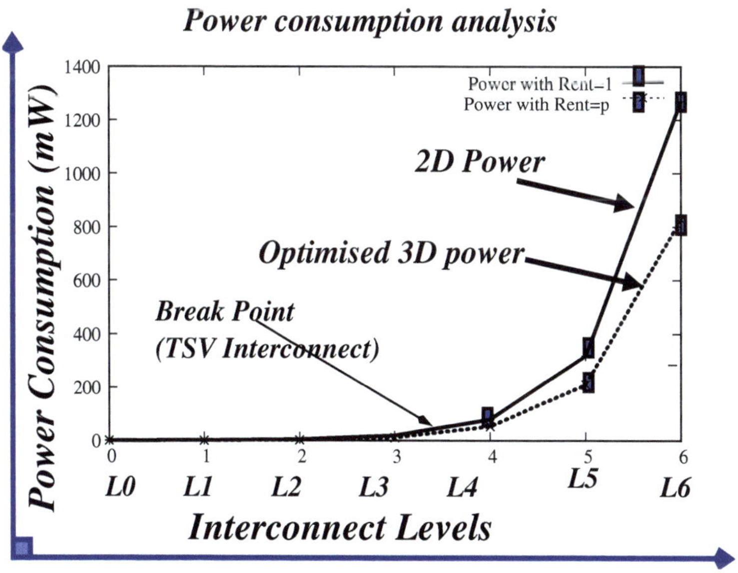

Fig. 6.10 Power consumption analysis of 3D Tree-based programmable interconnect network

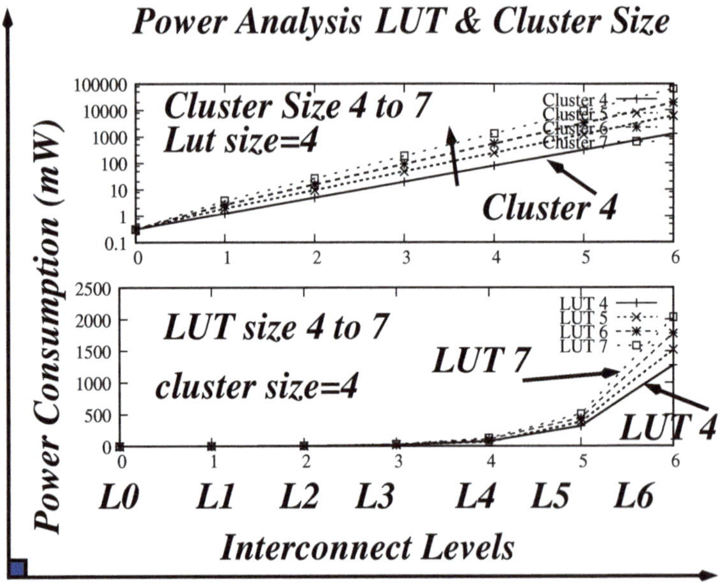

Fig. 6.11 Impact of cluster and LUT size on power consumption

of 3D chip. This doubles the total FPGA power for two-tier Mesh-based FPGA and this leads to pessimistic prediction of inter-layer temperature. While for Tree-based 3D FPGA, the power consumption of the dies in each tier is balanced through the optimization process of routing resources and TSV count. Figure 6.10 shows the interconnect power at different levels of the 3D Tree-based FPGA. The Rent's parameter based architecture optimization model shows 35.13 % reduction in total power consumption of 7 level Tree-based 3D interconnect network. This is very promising for FPGA architecture in terms of silicon area, since FPGA is an interconnect-dominated architecture and it is impossible to manufacture it with huge number of TSV and switches. Figure 6.11 presents the effect of LUT and cluster size on estimation of power consumption. The power consumption increased exponentially as LUT and cluster size increase due to exponential growth of switch size as the tree grows to higher levels. Considering the power consumption and performance results, LUT and cluster size equal 4 and 5 is the best architecture for manufacturing 3D FPGA. Nonetheless higher LUT and cluster size can be used where performance is considered to be the major design criterion.

6.10 Summary

Tree-based FPGA architecture is an old concept and due to numerous issues regarding speed and 2D physical design, the Tree-based architecture took the back seat compared to industry dominance of Mesh-based FPGA architectures. Due to advent 3D technology, many traditional design methods have been revisited to study those issues which hinters its march towards manufacturing and production. This chapter provides insights about how to resolve such issues and advantages compared to industrial Mesh-based FPGA architecture. A completed set of 3D CAD tools are developed to design and validate 3D Tree-based FPGA. The horizontal and vertical partitioning methodology based on design specification is a defining feature. An innovative multilevel network architecture and the issues associated with vertical and horizontal partitioning, TSV count optimization and its impact on design and manufacturing of 3D FPGAs are studied and presented. The study reveals the challenges in optimizing the physical design of 3D Tree-based FPGA and TSV management in a 3D stacked chip. The architecture level interconnect optimization model based on Rent's parameter shows minimal degradation in interconnect delay. Therefore the design and architecture optimization results presented in this chapter can serve as a robust foundation for the design and manufacturing of even more practical 3D re-configurable systems based on Tree-based FPGA architectures.

References

1. C. Ababei, P. Maidee, K. Bazargan, *Exploring Potential Benefits of 3D FPGA Integration, Field Programmable Logic and Application*, vol. 3203 (Springer, Berlin, 2004), pp. 874–880
2. K. Siozios, A. Bartzas, D. Soudris, Architecture level exploration of alternative schmes targeting 3D FPGAs: a software supported methodology. Int. J. Reconfigurable Comput. Article ID 764942, 18 (2008)
3. S. Gupta, M. Hilbert, S. Hong, R. Patti, *Techniques for Producing 3D ICs with High-Density Interconnect* (Tezzaron Semiconductor, Naperville, 2005)
4. C-I Chen, B-C Lee, J-D Huang, Architectural exploration of 3D FPGAs towards a better balance between area and delay, in *Design, Automation and Test in Europe Conference and Exhibition(DATE), IEEE*, Grenoble, France, 2001
5. K. Siozios, Vasilis F. Pavlidis, D. Soudris, A novel framework for exploring 3-D FPGAs with Heterogeneous interconnect fabric. ACM Trans. Reconfigurable Technol. Syst. **5**, 1–5 (2012)
6. B. Landman, R. Russo, On a pin versus block relationship for partitions of logic graphs. IEEE Trans. Comput. **20**(12), 1469–1479 (1971)
7. C. Ababei, H. Mogal, K. Bazargan, Three-dimensional place and route for FPGAs. IEEE Trans. Comput-Aided Des. Int. Circuits Syst. **25**(6), 1132–1140 (2006)
8. C. Ababei, Y. Feng, B. Goplen, H. Mogal, T. Zhang, K. Bazargan, S. Sapatnekar, Placement and routing for 3D integrated circuits, IEEE Des. Test **22**(6), 520–531 (2005)
9. A. Gayasen, V. Narayanan, M. Kandemir, A. Rahman, Designing a 3-D FPGA: switch box architecture and thermal issues. IEEE Trans. Very Large Scale Integr. (VLSI) Syst. **16**(7), 882–893 (2008)
10. Z. Marrakchi, H. Mrabet, U. Farooq, H. Mehrez, FPGA interconnect topologies exploration. Int. J. Reconfigurable Comput. **37**, 13 (2009)

11. U. Farooq, H. Parvez, Z. Marrakchi, H. Mehrez, A new heterogeneous Tree-based application specific FPGA and its comparison with mesh-based application specific FPGA. Elsevier Microprocess. Microsyst. J. Appl. Reconfigurable Comput. Special Issue **36**(8), 588–605 (2012)
12. V. Pangracious, Z. Marrakchi, E. Amouri, H. Meherez, Performance analysis and optimization of high density Tree-based 3D Multilevel FPGA, in *ARC11-2013, Reconfigurable Computing: Architectures, Tools and Applications Lecture Notes in Computer Science (7806)* (Springer, LA, 2013), pp. 197–209
13. A. DeHon, Unifying mesh- and tree-based programmable interconnect. IEEE Trans. Very Large Scale Integr. (VLSI) Syst. **12**(10), 1051–1065 (2004)
14. A. DeHon, R. Rubin, Design of FPGA interconnect for multilevel metallization. IEEE Trans. Very Large Scale Integr. (VLSI) Syst. **12**(10), 1038–1050 (2004)
15. J. Pistorius, M. Hutton, Placement rent exponent calculation methods, temporal behaviour and FPGA architecture evaluation, in *Proceedings of the International Workshop on System Level Interconnect Prediction* (Monterey, California, 2003), pp. 31–38
16. A. DeHon, Balancing interconnect and computation in a reconfigurable computing array (or, why you don't really want 100% LUT utilization), in *Proceedings of the 1999 ACM/SIGDA seventh international symposium on Field programmable gate arrays (FPGAs)* (Monterey, California, 1999), pp. 69–78
17. I. Koun, J. Rose, Measuring the gap between FPGAs and ASICs, in *Proceedings of the ACM/SIGDA International Symposium on Field programmable gate arrays FPGA'06* (Monterey, California, USA, 2006)
18. Z. Marrakchi, H. Mrabet, C. Masson, H. Mehrez, Mesh of tree: unifying mesh and MFPGA for better device performances. in *NOCS 2007* (2007), pp. 243–252
19. V. Pavlidis, E.G. Friedman, Interconnect-based design methodologies for three-dimensional integrated circuits, in *Proceedings of the IEEE*, pp 123–140, Jan 2009
20. L. McMurchie, C. Ebeling, PathFinder: a negotiation based performance driven router for FPGAs. in *ACM/SIGDA International Symposium on of Field Programmable Gate Arrays FPGA*, vol. 12, issue no 2, pp. 291–301, June 1995
21. M. Pathak, Y. Joon Lee, T. Moon, S. Kyu Lim, Through-silicon-via management during 3D physical design: when to add and how many?, in *IEEE-ICCAD, 2010*, pp 387–394 (2010)
22. Lip6 DSP Benchmarks SETI (2007), http://www-asim.lip6.fr/. 2007
23. Heterogeneous Benchmark Set II 2010. University of Toronto FPGA Research Group (2010), http://www.eecg.utoronto.ca/vpr/. 2010
24. Open Cores SET II 2010. Open Cores Benchmark set (2010). http://www.opencores.org/
25. U. Farooq, H. Parvez, Z. Marrakchi, H. Mehrez, Exploration of heterogeneous FPGA architectures. Int. J. Reconfigurable Comput. Special issue ReCoSoC 2010–2011, issue 2 (2011)
26. S. Yang, *Logic Synthesis and Optimization Benchmarks, Version 6.0.* (MCNC: Microelectronic Centre of North Carolina Research Triangle Park, North Carolina, 1991)
27. R. Le, S. Reda, R. Iris Bahar, High-performance, cost-effective heterogeneous 3D FPGA architectures. IEEE/ACM Int. GLSVLSI **2**, 218–229 (2011)
28. A. Marquart, V. Betz, J. Rose, Using cluster-based logic block and timing-driven packing to improve FPGA speed and density, in *Proceedings of the 1999 ACM/SIGDA Seventh International Symposium on Field Programmable Gate Arrays (FPGAs)* (Monterey, 1999), pp. 37–46
29. E. Ahmed, J. Rose, The effect of LUT and cluster size on deep-submicron FPGA performance and density. IEEE Trans. Very Large Scale Integr. (VLSI) Syst. **12**(3), 288–298 (2004)
30. M. Lin, A.E. Gamal, Y.-C. Lu, S. Wong, Performance benefits of monolithically stacked 3D FPGA, in *Proceedings of the 2006 ACM/SIGDA 14th International Symposium on Field programmable Gate Arrays* (Monterey, California, 2006), pp. 113–122, 22–24 Feb 2006
31. S. Simon Wong, A. El-Gamal, The prospect of 3D-IC, in *IEEE Custom Integrated Circuit Conference (CICC), IEEE* (San Jose, CA, 2009), pp. 445–448
32. F. Li, D. Chen, L. He, J. Cong, Architecture evaluation for power-efficient FPGAs, in *Proceedings of the ACM/SIGDA International Symposium on Field Programmable Gate Array*, pp. 175–184, Nov 2003

Chapter 7
Three-Dimensional Thermal Modeling: Tools and Methodologies

Abstract A 3D-IC system consists of disparate materials with considerably different thermal properties including semiconductor, metal, dielectric, and possibly polymer layers used for inter-plane bonding. Although the power consumption of these circuits is expected to decrease due to the considerably shorter interconnects, the power density increases since there is a greater number of devices per unit volume as compared to a 2D circuit. Heat transfer analysis in 3D-ICs is complicated by the presence of multiple heat sources and the introduction of new thermal resistances posed by inter-die materials including interface resistances whose values are not readily available. In this chapter we present a fast and accurate 3D thermal model developed for an n-tier 3D stacked Tree-based FPGA chip using 3D R-C mesh-based model. The design and implementation of 3D thermal analysis model depends on the characteristics materials and layers used to manufacture VLSI chips.

7.1 Introduction: Thermal Fundamentals and Challenges

Semiconductor analysts suggest that increasing power density due the exorbitant number of transistors and resulting difficulty in controlling on-chip temperature are some of the most critical challenge we face today to continue scaling of VLSI systems. Before we move into the details and challenges in 3D thermal modeling, some reviews and revisit of the two fundamental physical mechanism, heat generation and heat transfer, is needed in order to fully understand the topic of this chapter. Let us start with definition of temperature. Temperature can be defined in many ways. In a *macroscopic* standpoint the temperature can be defined as the property shared by two systems, initially at different states, after they have been placed in thermal contact and allowed to come to thermal equilibrium [18]. However in a *microscopic* standpoint: Temperature is directly proportional to the square of the mean molecular speed. As temperature increases, the molecules moves faster. From these definitions we can state that temperature is a state variable reflecting the level of the internal energy posses by a system.

V. Pangracious et al., *Three-Dimensional Design Methodologies for Tree-based FPGA Architecture*, Lecture Notes in Electrical Engineering 350, DOI 10.1007/978-3-319-19174-4_7

7.1.1 Heat Generation

The basic functional unit of CMOS integrated circuits is a MOS transistor. Each transistor can be turned on or off like a switch depending on the voltage difference between the gate and the source terminal. At a higher abstraction level, a number of MOS transistors are connected together in a particular topology to form a logic function, for example a logic block of FPGA and has its own input signals and output signals. Computation is the process to get the correct output voltage level for a given binary combination of the input signals. From basic circuit theory, we know that the total energy drawn from the power supply for this voltage transition is $C_L V_{dd}^2$, in which C_L is the load capacitance and V_{dd} is power supply voltage. However the actual energy stored in the circuit is only $\frac{1}{2} C_L V_{dd}^2$ and the other half is dissipated in the form of *Joule* heat in the inherent resistance of the circuit itself. Therefore every switching event as a result of computation draws some amount of energy from the power supply and this energy is eventually transformed into the heat dissipation.

7.1.2 Heat Transfer

All the generated heat in the integrated circuits must be removed or transferred to the ambient environment. Otherwise, the operating temperature will accumulate and cause malfunction and eventually the destruction of the system. Heat transfer is the transport of thermal energy from one region to another. In order for the heat transfer to occur, there must be a temperature difference between the two regions [4]. From the first law of thermodynamics, i.e. the conservation of energy, states that the heat given by the hot region has to be equal to the heat absorbed by the cold region. In addition, the second law of thermodynamics states that heat must be transferred from hot region to cold region. That means heat flows in the direction of decreasing temperature. There are three modes of heat transfer: conduction, convection and radiation [4]. Since we deal with semiconductor chips, conduction is major heat transfer mode considered in our model. The governing equation of heat conduction is the $Fourier's$ Law:

$$q = -k\frac{dT}{dx} \tag{7.1}$$

The Eq. 7.1 is the one dimensional form of the $Fourier's$ Law, where q is the heat flux (in W/m^2), k is the thermal conductivity of the material (in W/(m K)). The heat flux q (is the flow of heat per unite area per unit time), at a point in a medium is directly proportional to the temperature gradient at that point. The minus sign indicate that heat flows in the direction of decreasing temperature. If we substitute $q = Q/A$, where Q is the heat transfer rate, A is the heat conducting area, and assume L is the length of the conductor the Eq. 7.1 becomes

$$Q = -kA\frac{\Delta T}{L} \tag{7.2}$$

and if we define thermal conductance $g_{th} = Q/\Delta T$, the heat transfer rate divided by temperature drop, we get

$$g_{th} = Q/\Delta T = \frac{kA}{L} \tag{7.3}$$

and the Eq. 7.3 resembles the well known Ohm's Law in the electrical circuit theory:

$$\sigma = I/\Delta V = \frac{A}{\rho L} \tag{7.4}$$

Therefore, we have the following interesting duality between electrical and thermal phenomena, thermal conductance k versus electrical conductance $1/\rho$; temperature difference ΔT versus voltage difference ΔV; heat transfer rate Q versus electrical current I and thermal resistance $R_{th} = 1/g_{th}$ versus electrical resistance $R = 1/\sigma$. Heat conduction is also a transient process, a more general equation which also considers time in the heat diffusion equation:

$$\rho c_p \frac{\partial T(x, y, z, t)}{\partial t} = \nabla .[k(x, y, z, t)\nabla T(x, y, z, t)] + g(x, y, z, t) \tag{7.5}$$

where ρ is the density of the material (kg/m^3), not the electrical resistivity, and g is the volume power density of the heat source(s) (W/m^3), c_p is the specific heat (J/(kg °C)). While thermal conductivity k actually is a function of location and temperature, we can assume it is isotropic and temperature independent, as this is mostly true for the materials and the temperature range that we are interested in. For the steady-state case, the $\partial T/\partial t$ term in Eq. 7.5 becomes zero. We can verify that at steady state, the one-dimensional form of the heat diffusion equation can be reduced to the Fouriers Law Eq. 7.1. If we assume both g and k are constant, Eq. 7.5 can be rearranged to the following from if we write it in one-dimensional form and integrate both sides by the integral variable x from 0 to L (the length of the material), and notice that $g.L = Q/A = q$, the heat flux, we get

$$(\rho c_p AL)\frac{dT(t)}{dt} = kA\frac{\Delta T(t)}{L} + Q \tag{7.6}$$

The first term of the right-hand side of Eq. 7.6 is the teat transfered through the thermal resistance R_{th}, similar to that in Eq. 7.3. Note that $\Delta T = T_1 - T_2$, and move this term to the other side of the equation, we get

$$C_{th}\frac{dT(t)}{dt} + \frac{T_1 - T_2}{R_{th}} = Q \tag{7.7}$$

Table 7.1 Duality between thermal and electrical properties

Thermal quantity	Unit	Electrical quantity	Unit
Q Heat transfer rate, power	W	I, Current	A
T, Temperature difference	K	V , Voltage difference	V
R_{th}, Thermal resistance	K/W	R, Electrical resistance	Ω
C_{th}, Thermal capacitance	J/K	C, Electrical capacitance	F

where $C_{th} = \rho c_p AL = c_p \rho V$ is defined as thermal capacitance or thermal mass, V is the volume of the material. From the electrical circuit theories, we know that $C\frac{dV(t)}{dt} = i_c(t)$, meaning the current flow through an electrical capacitor equals the product of its capacitance and the first derivative of the voltage difference across it. This exactly resembles the first term on the left-hand side of Eq. 7.7. This is the reason we define C_{th} as thermal capacitance. Thermal capacitance describes the heat absorbing capability of a material, while electrical capacitance describes the ability of accumulating electrical charges of a material. Equation 7.7 states that the total heat flowing through the material is equal to the sum of the heat flowing through the thermal capacitance (the AC component) and the heat flowing through the thermal resistance (the DC component). Table 7.1 summarizes the duality between thermal and electrical phenomena. We use this duality to derive the compact thermal resistance and capacitance network to formulate the thermal model parameters for 3D FPGAs. The heat conduction equation derived above are only applicable to the macroscopic world and will not provide accurate temperature estimation, if the minimum feature size scales below 300 nm, because when the dimension scales in the sub-nano regime, quantum effects needs to be considered since the mean free path for phonon-phonon scattering is $\approx$300 nm for silicon. In this case it is better to consider Boltzmann Transport Equation (BTE) to accurately model thermal effects at nanoscale transistor level. In this section, we will be using block level thermal simulations for 3D FPGAs.

The operating temperature is related to power and power density according to Eq. 7.3. However temperature is not simply proportional to the power consumption, neither the power density. there are other factors that significantly impact temperature distribution in space and time that are also needed to be taken care of. These factors include heat spreading and temporal and spatial temperature filtering effects. Equation 7.3 does not account these effects and therefore temperature must be modeled directly in order to perform accurate thermal analysis during the design process. Heat spreading happens when heat transferred from a small surface area to a large one. Temperature filtering happens in the time domain where the long thermal time constant of the silicon and package tends ti filter out fast changes (high frequency component) in power and power density. Temperature filtering can also happen spatially where the power and power density change over a small dimension (high spatial frequency). All these effects can be modeled by solving the heat diffusion Eq. 7.5. But directly solving the 3D partial differential equation presented in Eq. 7.5 is a daunting task, if not impossible, without simplifications and numerical techniques. The thermal modeling methodology developed for Tree-based FPGAs provides an efficient and accurate way to construct compact thermal *R*-*C* networks to simplify the heat diffusion equation.

7.1.3 State of the Art: Thermal Modeling

The increase in internal temperature is a growing concern in modern FPGA designs. Recent articles on thermal management from leading FPGA manufactures [13, 19] clearly indicate the growing importance of thermal issues in FPGA designs. The logic density improved more than 3 times [20] and frequency of operation of all latest FPGA designs due to improvement in manufacturing technology, circuit design, architecture and tool development. However this may also lead to increase in power and power densities, which manifests itself in the form on high temperatures. Many design level techniques has been adopted to remove and balance heat generated by the design. Thermal-aware floorplanning methods tries to reduce the hotspots on the die by distributing the temperature uniformly [15, 16]. A thermal aware placement method based on partition-driven algorithm proposed in [9] for standard cell placement. Several researchers have developed thermal modeling tools to estimate the die temperature. Among them *HotSpot* [10] is an micro-architectural level thermal simulator, which can perform transient as well as steady state temperature estimation. HotSpot provides flexibility to set several package and die parameters, such as the heat spreader thickness, package-to-air thermal resistance and substrate thickness. From above discussion, we see that it is very important to be able to estimate temperature at different granularities and at different design stages, especially early in the design flow. The estimated temperature can then be used to perform power, performance, and reliability analysis, together with placement, packaging design, etc. As a result, all the decisions use temperature as a guideline and the design is intrinsically thermally optimized and free from thermal limitations.

7.2 3D Thermal Modeling

The traditional 2D chip design and fabrication technology facing challenges in integrating the exponentially growing number of transistors in a single chip. The wire delay and power consumption are increasing dramatically and achieving interconnect design closure in modern design is becoming a major challenge. The 3D technology results in smaller footprint in each layer and shorter vertical wires that are implemented using Through Silicon Vias (TSVs) across the layers. Despite the advantages of 3D-ICs over 2D-ICs, thermal effects are expected to be significantly exacerbated in 3D-ICs due to higher power density and greater thermal resistance of the insulating dielectric, which leads to higher junction temperatures. A 3D system consists of disparate materials with considerably different thermal properties including semiconductor, metal, dielectric, and polymer layers used for plane bonding. Although the power consumption of these circuits is expected to decrease due to the shorter interconnects (TSVs), the power density increases since there is a greater number of devices per unite volume as compared to 2D design [5, 7]. These issues make the modern sub-nano technology designs more complex and calls for the integration of efficient heat removal mechanism at various stages of modern circuits designs flow, such as synthesis, floorplanning, placement and routing to maintain the temperature of circuit designs within the specified limits. The power consumption of 3D-ICs is expected to decrease due to the interconnect length reduction. However the power density increases since the distance between the devices decreases per unit volume as compared to a 2D layout. Consequently, the temperature rises among the layers placed far from the heat sink of the package, resulting in performance and reliability degradation. Thermal analysis of modern FPGA architectures is essential, as the power dissipation and leakage can be expected to increase as we scale down design and manufacturing technology below sub-nanometer regime. The absence of effective heat removal solutions may lead to performance and reliability degradation in 3D-ICs. It is essential to develop design and hardware techniques to allow effective thermal conduction among tiers of 3D system to maintain the performance of the multi-tier 3D FPGAs. The die temperature must be control rise in internal temperature, because it has huge impact on circuit timing, leakage power, package design and also lifetime of the device. A high internal temperature dramatically impact the performance of the system.

The increase in internal temperature can impede device performance and reliability of 3D-ICs. Thus it essential to develop 3D-specific design tools which includes thermal design techniques. This will enable the designers to address thermal issues and generate reliable and high performance 3D designs. Two key elements are required to establish a successful thermal management strategy; a thermal model, to characterize the thermal behavior of a circuit, and design techniques that alleviate thermal gradients among the physical tiers of a 3D stack while maintaining the operating temperature within acceptable levels. Thermal design techniques can be classified into two categories: heat removal strategies that improve the thermal profile of the 3D chip without requiring any redundant interconnect resources for thermal management, such as using the signal, power and ground TSVs for heat transfer and those methodologies that are an integral part of a more aggressive thermal policies such as that utilizing vertical thermal TSVs and wires and other more expensive liquid cooling methods [6, 28], sacrificing some of the design objectives. These TSVs are typically called thermal or dummy TSVs [2, 28] to emphasize the objective of transferring heat rather than providing signal communication for circuit located at different tiers of the 3D chip. Thermal wires corresponds to those horizontal wires that connect regions with different thermal via densities through thermal TSVs. This is done to remove or balance heat uniformly inside the 3D stack.

Recently, some researchers also proposed solutions and methods for evaluation of thermal issues in 3D-ICs. Numerical thermal simulations have been carried out to convert power dissipation distribution into a temperature distribution in a 3D-ICs.

The development of a fundamental analytical model for heat transport in 3D integrated circuits is highly desirable to provide a framework in which to analyze the general problem of heat dissipation in 3D ICs and will offer simple thermal design guidelines. A thermal-driven floorplanning tool for 3D-ICs proposed in [14] and a thermal-driven for 3D standard cells dedicated to ASICs proposed in [8]. Another work [24] propose analytical and finite-element model of heat transfer in 3D electronic circuits and use this model to analyze the impact of various geometric parameters and thermophysical properties like through silicon vias, inter-die bonding layers etc. on thermal performance of 3D-ICs. As discussed in Chap. 5, TSV is a key component of 3D integration technology. A 3D-IC thermal inter-tier thermal control using TSV based nano-structure is proposed in [27] to effectively optimize the thermal profile of 3D-ICs. A recent FPGA thermal analysis study proposed alternative organization for a 2-dimensional FPGA to reduce the intra-die thermal variations [21, 25] and demonstrated a peak temperature reduction 6 °C using a fully utilized $Vertex—4FX100$. The thermal organization indeed has a greater impact, since the thermal variation in 3D FPGA is larger. A compact 3D thermal model for liquid cooling for fast thermal simulation of 3D-ICs with inter-tier micro-channel cooling called *3D-ICE* is proposed in [28]. A 3D FPGA thermal simulator based on *3D-ICE* model is implemented to model and validate different heat transfer methods developed of 3D Tree-based FPGAs.

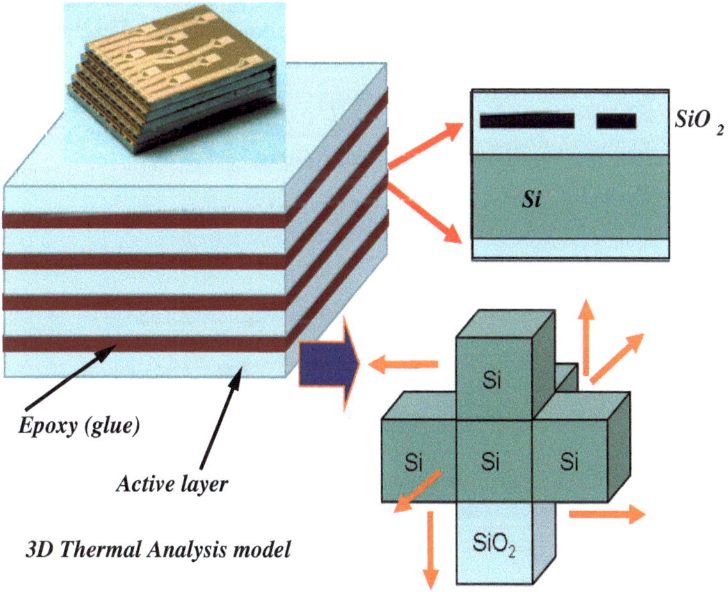

Fig. 7.1 Illustration of how 3D-ICs are structured into thermal units for 3D thermal analysis

7.3 Heat Transfer in 3D-ICs

Heat transfer analysis in 3D-ICs is complicated by the presence of multiple heat sources and the introduction of new thermal resistances posed by inter-die materials including interface resistances whose values are not readily available. In this section we present a 3D thermal model developed for a n-tier 3D stack FPGA chip using 3D $R-C$ mesh-based model from *3D-ICE*. The design and implementation of accurate and fast thermal analysis model depends on the characteristics materials and layers used to manufacture VLSI chips. There are several heat fluxes associated with each active layer: (1) in-plane heat transfer due to thermal conductivity of silicon; (2) vertical heat transfer between layers through adhesive layers, BEOL (back-end-of-line) metal dielectric layer and TSV material and location. The heat transfer in an adhesive layer is considered to be perpendicular to the device plane and no heat flux occurs along the adhesive layer due to its low thermal conductivity. The heat transfer in TSVs is considered as one dimensional and perpendicular to the device plane. Heat fluxes for top and bottom layers are also defined by heat transfer through a heat sink and packaging. The heat flow inside the 3D stack is diffusive in nature and hence it can be modeled by its equivalence to an electronic RC circuit [10, 12, 24] as established in Sect. 7.1.2. This is done by first dividing the entire chip structure into small cubical thermal cells as illustrated in Fig. 7.1. Each cell is then modeled as a node containing six resistance that represent the conduction of heat in all the six directions (top, bottom, north, south, east and west), and a capacitance that represent the heat storage inside the cell as shown in Fig. 7.2. The conductance of each resistor and capacitance of the thermal cell are calculated as follows.

$$g_{top/bottom} = k_{th} \times \frac{l \times w}{(h/2)} \tag{7.8}$$

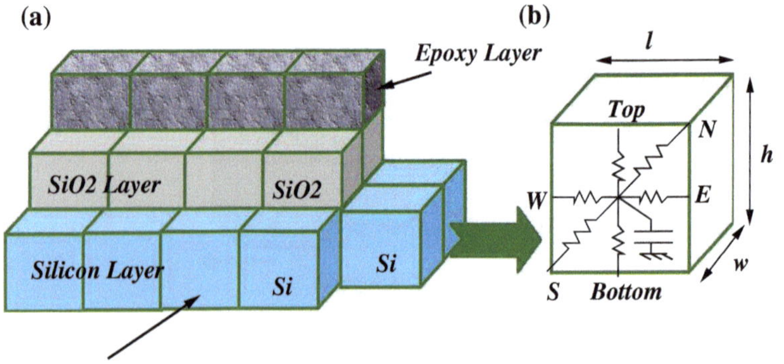

Fig. 7.2 **a** The unitary thermal calls of the 3D stack. **b** Equivalent RC circuit of single cells

$$g_{north/south} = k_{th} \times \frac{l \times h}{(w/2)} \tag{7.9}$$

$$g_{east/west} = k_{th} \times \frac{w \times h}{(l/2)} \tag{7.10}$$

$$C_{top} = SC_{th} \times (l \times w \times h) \tag{7.11}$$

Here the subscripts *top, east, south north* etc., indicate the direction of conduction, k_{th} and SC_{th} are the thermal conductivity and specific heat capacity per volume unite of the material, respectively. The entire circuit is grounded to the ambient temperature at the top and the side boundaries of the 3D stack through resistance, which represent the thermal resistance from the chip to the air ambient. The behaviour of the resulting RC circuit can be described using a set of first-order differential equations via nodel analysis [1] as follows.

$$B.U(t) = G.X(t) + C.X(t) \tag{7.12}$$

where $X(t)$ is the vector of cell temperature of the circuit at time t, G and C are the conductance and capacitance matrices of the circuit, U(t) is the vector of the input heat sources (in this case power sources) for example a logic block or a switch block in an FPGA and B is a selection matrix G and C present a sparse block-tridiagonal and diagonal structure, respectively, due to the characteristics and definition of the thermal problem. In addition, G and $U(t)$ are functions of the cell temperatures $X(t)$, making the behavior of the circuit non-linear. This is because of the temperature-dependent thermal conductivity of silicon and the temperature dependent electrical resistance of the metal interconnects used in BOEL process respectively. In our thermal model, a first-order dependence of these parameters on temperatures around 300 K is assumed. Some of these parameters are presented in Table 7.2 [22].

Table 7.2 Thermal properties of materials

Thermal properties of materials	Values
Silicon thermal conductivity	$295–0.491 \times T$ W/m K
Silicon specific heat	1.659×10^6 J/m^3 K
SiO_2 thermal conductivity	1.38 W/mK
SiO_2 Specfic heat	4.180×10^6 J/m^3 K
Aluminum electrical resistivity	$2.82 \times 10^{-8}(1 + 0.0039\Delta T)\Omega$m
	$\Delta T = T - 293.15$ K
Copper (C_u) thermal conductivity	386.01 W/(m K)
Tungsten (W) thermal conductivity	162.714 W/(m K)
Heat sink heat transfer	2.0 W/kK
Package heat transfer	0.2 W/kK

The DC solution for the circuit can be found be solving the corresponding steady state Eq. 7.13.

$$GX = BU \tag{7.13}$$

The above set of equations are solved by the inversion of the matrix G using the sparse LU decomposition method [3]. Because of the non-linearity of the circuit and the input power sources, these equations were solved repeatedly, by updating the matrices after each iteration of solving, until convergence is reached. Many researchers and industrial partners proposed fabrication based solutions for the thermal management in 3D integrated circuits. Thermal through silicon vias (TTSVs) have a prominent place among these solutions. It is important to reduce the temperature gradient between various parts of the chip because variations in operating temperature affects the performance of the chip, leading to timing errors and chip failures. Moreover thermal gradients have been observed as a determinant negative factor on system reliability.

7.4 3D Tree-Based FPGA Thermal Analysis Model

Thermal issues in 2D and 3D FPGAs are relatively unexplored. In our 3D thermal experiments, we considered all FPGA blocks such as LBs, HBs and SBs of the Tree-based interconnect levels range from 0 to 6. To analyze the thermal variation in 2D and 3D FPGA, we used a two-tier 3D Tree-based FPGA design. The tier 1 consists of LBs and local SB interconnects up to level 3 and tier 0 consists of HBs and SBs of higher level interconnection network of the Tree architecture. The structure of two-tier experimental 3D stack is presented in Fig. 7.3. In general the F2F (face-to-face) 3D stacking configuration is used for two-tier 3D structure with help of solder bumps. However this limits the number of stacked tires to two [29] or to multiple stacks of F2F structure. In this work, we use a F2B (face-to-back) stacking configuration, since we expect to design and build multiple tier high density 3D Tree-based FPGAs, as the Tree grows to higher levels. For Mesh-based 3D FPGA, the researchers assumed the same power for individual blocks that was stacked in a face-to-back manner [26]. This doubles the total power consumption of two-tier Mesh-based FPGA compared to monolithic 2D FPGA and also leads to pessimistic temperature estimation, because the total power consumption is expected to be lower for 3D FPGAs, as the wire length of the interconnect network reduces, while we stack multiple tiers. For 3D Tree-based FPGA the total power consumption is redistributed across the blocks that are assigned in tiers 0 and 1 and this model provides an accurate estimation of inter-tier temperature of the two-tier Tree-based FPGA structure. Figure 7.3 shows the organization of LBs and local interconnects in tier 1 and the higher level interconnect network of the Tree in tier 0. To distribute the heat generated in 3D Tree-based FPGA, we developed two different heat transfer methodologies. (1) based the thermal-aware design techniques and (2) based on more

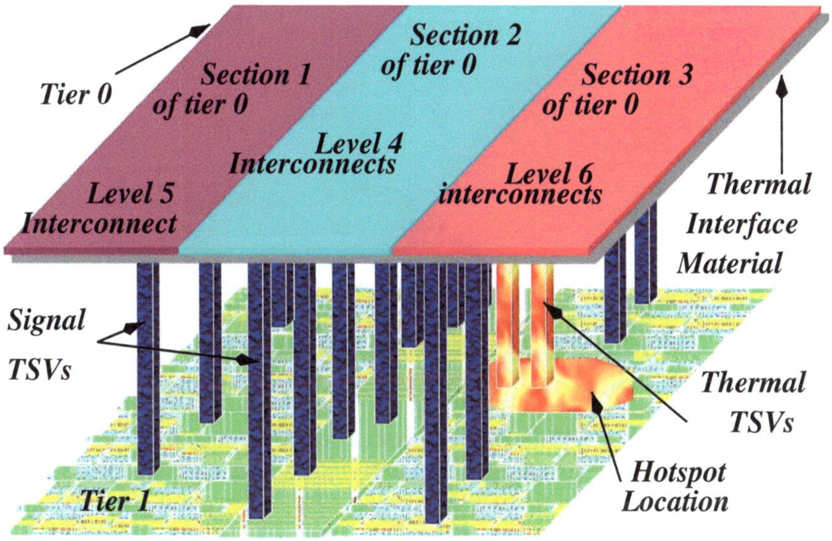

Fig. 7.3 2 layer 3D tree-based FPGA experimental structure used for thermal analysis

aggressive heat transfer policy using hardware based techniques using additional TSVs and metal wires.

7.4.1 3D Thermal Aware Design Techniques

The goal is optimize the temperature generated inside the chip by distributing the power sources equally across the active layer of the 3D FPGA design. As we discussed in Chap. 8, the multilevel Butterfly-Fat-Tree (BFT)-based programmable interconnect network is partitioned at a particular level called the *break-point* level and interconnected using TSVs. The 2D Tree-based FPGA design is partitioned based on design specification (*horizontal* or *vertical*) to form a two-tier 3D Tree-based FPGA. For *vertical* partitioning methodology, the design is divided into two equal section in terms of area and power consumption. In this case the placement of power sources on both layers of the 3D chip are balanced. However in the case of *horizontal* partitioning methodology, the interconnect network is partitioned at a certain level called the *break-point* based on the network delay optimization and this partitioning method gives only little room for thermal optimization. To understand the thermal distribution of this design, we used the thermal-aware floorplane tool [16] along integrated in our physical design flow. The thermal-driven floorplanning tool is configured with Global Foundries 130 nm technology node. This tool is configured to estimate temperature of the blocks floorplan of the two-tier Tree-based FPGA chip based on its power consumption and connectivity ratio among them. The floorplan

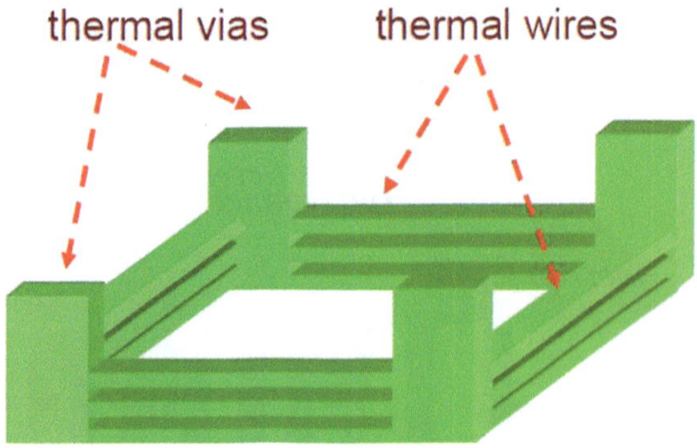

Fig. 7.4 Thermal network using TSVs and *horizontal* wires to transfer heat from hotspot to coldspot

tool takes a list of functional blocks, areas, aspect ratios, connectivity ratio between the blocks and power consumption of each functional blocks as inputs. For example, in the case of horizontal partitioning, we have have two floorplans: the first floorplan consists of LBs, and local interconnections up to *level 3* of the Tree-based FPGA and the second floor plan consists of programmable interconnect levels *levels 4, 5, 6*, hard-blocks (HBs) and I/Os.

The floorplan tool generates thermal estimations of local and global metal layer. For this study the communication is realized with Through Silicon Via (TSV) using TSV Tezzarion TSV technology [17] and electrical characterization of TSV is performed using the approach presented in [23]. One important aspect of thermal-aware floorplanner is the trade-off between temperature and performance. We used the wire delay model associated with floorplanner to optimize the wire length. However the floorplan solution is always a trade-off between temperature and wire delay of the blocks used in simulation. To manage this trade-off, we have taken steps during design phase to make sure the placement of high power blocks do not lead to hotspots

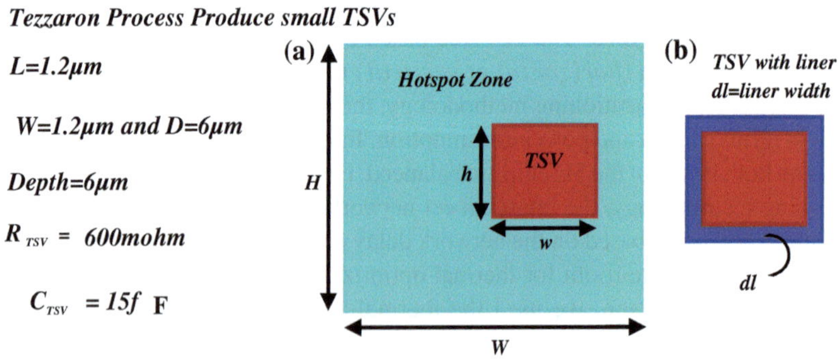

Fig. 7.5 **a** Top view of hotspot zone with TTSV placement, **b** TTSV with liner *dl*

without compromising on design performance. The floorplan tool is augmented with flexibility to create horizontal or vertical break-points in the BFT based interconnect network according to the 3D Tree-based FPGA design specifications. However by stacking multiple active tires and increasing logic density, it become more difficult to remove the inter-tier heat. Hotspot power dissipation results in significantly higher temperatures in 3D stacked chips compared to the same power dissipation in single 2D chips. Our study shows that there is a significant temperature gradient across the stacked dies for the two-tier Tree-based FPGA structures. The reason for the increase in temperature is due to the reduced thermal spreading in the thinned dies on the one hand, and to the use of low thermal conductivity adhesives on the other hand. Therefor a detailed thermal analysis at the design stage is required.

$$k_{eff} = k_{cu}.(TSV_{Area}) + K_{th}.(Level_{BPArea} - TSV_{Area}) \tag{7.14}$$

The floorplan tool uses 3D resistance mesh-based thermal model presented in Sect. 7.3 to extract the thermal profile of the floorplans of the two-tier 3D Tree-based FPGA. The 3D Thermal resistance mesh based multi-tier thermal model for Tree-based FPGA consider the spatial distribution of signal TSVs and PDN networks (power delivery) to control the heat transfer among different module in the multi-tier chip. In a 3D circuit, thousands of TSVs may co-exist to deliver the power and signal. These TSVs are normally filled with metal such as copper (Cu) or Tungsten which has a higher thermal conductivity than silicon and they can have a significant influence on the steady state temperature of 3D-ICs. Such influence is due to the change of thermal conductivity in the heat transfer path when a TSV is inserted and is highly dependent on the TSVs size and position. The thermal model also consider the impact of TSVs material (Cu,Tungsten or doped Poly-silicon) while estimating the temperature profile. The effective thermal conductivity of active and passive layers in 3D stacked chip is calculated by Eq. 7.14. The k_{cu} and K_{th} are the thermal conductivity of copper and silicon active layer. The heat transfer take place on those locations where Cu TSVs are placed. Using this module, the inter-layer heat transfer and thermal profile of 3D FPGA is modeled and analyzed.

7.4.2 TSV Aware Thermal Control

The major challenge in thermal management and development of 3D thermal analysis tools is the complexity of heterogeneous structure of different dies and poorly understood thermal behavior of TSVs and adhesive layers. In the previous Sect. 7.4 we discussed how to address the variations in the heat transfer characteristic of adhesive materials used for manufacture 3D-ICs to accurately estimate the internal hotspot temperature. In this section we study the impact of the distribution and total number of TSV on thermal performance using two-tier 3D Tree-based FPGA. In addition to this, we also implemented special zones inside the die to monitor the evolution of hotspot temperature. The special zone are located and placed based on initial thermal profile of the 3D chip. The aim is to implement a 3D thermal net using TSVs and horizontal wires to transfer the heat across the chip. To achieve this we use thermal-TSVs (TTSVs) and horizontal metal wires to form a 3D heat transfer network as illustrated in Fig. 7.4. The placement of additional thermal-TSVs for vertical heat transfer is a limited design technique to transfer inter-tier heat, because on one hand it increases the area and cost of the chip and on the other hand it also increase design complexity. However we found the 3D thermal net is an effective heat transfer technique based on TSVs. In this section we discuss the implementation of a fast and accurate steady state thermal simulator using TSV-based thermal control for 3D-ICs. Thermal conductance between adjacent cells are then calculated by considering the effect of TSVs. At steady state, the thermal balance equation holds for every grid cell as follows.

$$\sum (T_i - T_{i,adj}) \times g_{i,adg} = P_i \tag{7.15}$$

where T_i and $T_{i,adj}$ denote the temperatures of the ith grid cell and its adjacent grid cell, respectively. $g_{i,adj}$ represents the thermal conductance between the two cells, and P_i is the heat generated by the ith grid cell. From Eq. 7.15, it is obvious that an increasing P_i, in order to maintain T_i in an acceptable level, the thermal conductance has to be increased and this require a more advanced and expensive materials and in some case we may not have many choices than introducing addition device with high thermal conductance such as TTSVs. This is definitely a big concern in term of cost increase of the entire design. The matrix equation of thermal equilibrium can be solved by the sparse solver LU [3] with the extracted power consumption from two-tier FPGA design. In this work we used Tezzaron TSV technology, which produce very small squarish TSVs as illustrated in Fig. 7.5. The thermal conductance between two hotspot zones without TSVs is represented in Eq. 7.16.

$$g_{ij} = \frac{kA}{dx} \tag{7.16}$$

where k represent thermal conductivity, A is the cross-sectional area through which the heat flux passes and dx is the length of heat transfer path. When a TTSV is placed as shown in Fig. 7.5, the equivalent thermal conductance x, y, and z direction can be computed using the Eqs. 7.17, 7.18 and 7.19.

$$g_x = \left(1 + \frac{a_{tsv}(\beta - 1)}{A_{HZ}(1 + (1 - a_{tsv}/hW)(\beta - 1))}\right) \times g_{x0} \tag{7.17}$$

$$g_y = \left(1 + \frac{a_{tsv}(\beta - 1)}{A_{HZ}(1 + (1 - a_{tsv}/wH)(\beta - 1))}\right) \times g_{y0} \tag{7.18}$$

$$g_z = \left(1 + \frac{a_{tsv}(\beta - 1)}{WH}\right) \times g_{z0} \tag{7.19}$$

where $a_{tsv} = wh$ is the TTSV area, $A_{HZ} = WH$ is the hotspot zone area, $g_{x0} = \frac{k_{HZ} \times HL}{W}$, $g_{y0} = \frac{k_{HZ} \times WL}{H}$ and $g_{z0} = \frac{k_{HZ} \times WH}{L}$ are the thermal conductance of the Hotspot zone without TTSV in x, y, and z direction. L is the depth of the Hotspot zone. $\beta = k_{tsv}/k_{HZ}$ is the ratio of thermal conductance between TTSV and Hotspot zone. In the case of multiple TTSVs placed inside Hotpsot zone, we employ superposition principle. In 3D TSV technology implementation, TSVs are normally surrounded by a thin layer of less thermally conductive dielectrical liner for insulation purpose, as illustrated in Fig. 7.5. To address the thermal isolation of dielectrics, the equivalent thermal conductivity of TSV with liner is defined as follows.

$$k_{tsv,xy} = \left(1 + \frac{d_m^2(\gamma - 1)}{(d_m + 2d_l\gamma)}\right) \times k_{liner} \tag{7.20}$$

$$k_{tsv,z} = \left(1 + \frac{d_m^2(\gamma - 1)}{(d_m + 2d_l)^2}\right) \times k_{liner} \tag{7.21}$$

whre d_m is the side length of the TSV metal core, d_l is the liner thickness and $\gamma = k_{metal}/k_{liner}$ is the ratio of thermal conductivity between metal and liner. To calculate the thermal conductances between all the solid zones with TSV effects, thermal conductances without TSV are first obtained according to (7.15). The sizes and positions of TTSVs are then imported to determine their overlaps with all hotspot and non-Hotspot zones. Based on the equivalent thermal conductivity of every TTSV and TSV, their contributions to thermal conductance are then added up to the overlapped floorplan according to Eqs. 7.17, 7.18 and 7.19. The Hotspot zones with additional TTSVs gives flexibility to place TTSVs without compromising design requirements. The TTSVs are normally used to control the inter-layer temperature of the 3D chip [27]. Though it is not a cost-effective solution in terms of design and manufacturing, used in certain cases to transfer the inter-layer heat efficiently from a 3D chip. To experiment the impact of signal-TSVs and TTSVs we designed special Hotspot zones in the tier 0 based on the initial steady state thermal analysis presented in Fig. 7.9.

7.5 3D FPGA Thermal Modeling: Capabilities

The main features of the 3D thermal model implemented for thermal analysis of 3D Tree-based FPGA are as follows

- Accurate and efficient:
  1. Non-linear or impact on reliability and other design characteristics.
  2. Speed necessary for efficient 3D design exploration.
- Continuous heat flow analysis.

1. Capture geometrical characteristics of the devices from layout.
2. Possibility to explore complex 3D package structure.

- Model interface:

1. Input: Power model of tier devices, geometrical properties and placement.
2. Output: Temperature of tier devices at run-time.

- 3D Thermal Circuit:

1. Heat Flow → Electrical current; Temperature → Voltage
2. Metal and Silicon layers composed of elementary thermal blocks

- Simulation capabilities:

1. Extensible set of tiers in 3D stack
2. Simulate up to 20 tiers and heat spreader
3. Pre-defined active and passive layers: Silicon, copper, glue overmould, interposer with and without TSVs, microbumps.
4. Configurable number of cells and iterations per tiers using grid model.
5. Capture TSV characteristics and location. Effective thermal conductivity method is adopted.
6. Special heat transfer zones can be setup and number of TTSVs can be specified.

The 3D thermal model tool is design to run 2D and 3D thermal simulations separately. To run 3D simulation, the user needs to specify layer information file, which contains information about each tier in the 3D FPGA chip such as specific heat capacity in J/(m^3 K), resistivity in (m-K)/W, thickness in μm, heat transfer direction and floorplan files.

7.6 3D FPGA Thermal Modeling: Simulation Results

As discussed in Sect. 7.4, in our 3D thermal analysis experiments, we considered all FPGA blocks such as LBs, HBs and interconnect level SBs range from 0 to 6 of the tree-based interconnect structure. To analyze the thermal variation in 2D and 3D Tree-based FPGA, we use the physical design parameters of two-tier 3D Tree-based FPGA design as shown in Fig. 7.3 to generate the thermal analysis floorplan. The tier 1 design consists of LBs and local interconnects up to level 3 and tier 0 consists of I/O, HBs and higher level interconnection network (SBs) of the Tree architecture. The thermal analysis structure is configured using face-to-back 3D integration method. The area and power consumption of each blocks used in simulation are presented in Table 7.3. The floorplan used for simulation is presented in Fig. 7.6. The Fig. 7.6 presents the thermal profile of tier 1 of 3D Tree-based FPGA generated using 3D thermal model without using thermal design technique described in Sect. 7.4.1. The estimated peak temperature increased from 82 °C (2D temperature) to 111 °C for the two-tier 3D FPGA and average temperature is 105 °C. However by using the effective

Table 7.3 7 Level tree FPGA layout with 16k CLBs

Tree Levels = 7, Arity = 4, Arch = 4 × 4 × 4 × 4 × 4 × 4 × 4				
Tree-based FPGA block type	3D chip # layer	Rent = "p" Avg (21 MCNC)	Block Area (μm^2)	Power (mW)
CLB unit	1		93,635,273	0.15
Switch box Level 0	1	0.65	2412	0.25
Switch box Level 1	1	0.54	10,800	0.83
Switch box Level 2	1	0.66	37,496	2.70
Switch box Level 3	1	0.55	232,128	13.13
Break-point	Horizontal partitioning: TSV count = 45,056 Area = 540,672 μm^2			
Switch box Level 4	2	0.63	6,072,770	47.46
Switch box Level 5	2	0.64	41,553,499	214.05
Switch box Level 6	2	0.65	42,139,683	838.97

thermal conduction model, we were able to estimate the accurate value of hotspot peak and average temperature. Figure 7.7 shows the thermal profile of tier 1 using thermal design and effective thermal conductance method. In F2B type 3D design the TSVs pass from the metal 6 (M6) of tier 1 to metal 1 (M1) of the tier 0 of the 3D chip. In this case all TSVs will pass through the substrate of tier 0 which is thin down to TSVs. In our design it is only 6 μm. The estimated peak and average temperature reduced to 99 and 95 °C respectively. The estimated temperature values of using thermal design technique shows the need to consider the effect conductivity of all materials used in the chip design and manufactures. This increases the complexity of 3D thermal model. The temperature analysis of the two layer 3D Tree-based FPGA is presented in Fig. 7.8. The hotspot temperature estimated is still much higher than the acceptable level. Next section provides experimental analysis of hardware-based heat transfer in 3D designs.

The 3D thermal model has been improved with capabilities such as implementation special zones based on the initial thermal profile. Based on hotspot location and temperature a special hotspot zone can be designed and place additional Thermal TSVs to transfer heat as described in Sect. 7.4.2. With inclusion of TTSV-based zones, the thermal model considers the impact of spatial distribution of signal-TSV and Hotspot zones including TTSvs to compute the thermal profile of the 3D Tree-based FPGA chip. Figure 7.9 shows the new thermal zones generated in our floorplan to transfer heat from hotspot to other coldspots inside the chip. As illustrated in Fig. 7.9, the zones 2, 3, 4 and 5 are specifically meant for the placement of TTSVs and other zones were used to place horizontal metal wires to transfer heat generated at the hotspot. The area and size of hotspot special zone and TTSV count can be decided based on the thermal profile and design requirements. The thermal-TSV based heat transfer method increases the total area of the chip. To minimize the impact on area and design performance, we need to optimize the hotspot area and TTSV count. Figure 7.10 presents the two-tier floorplan and TSV distribution styles used in the

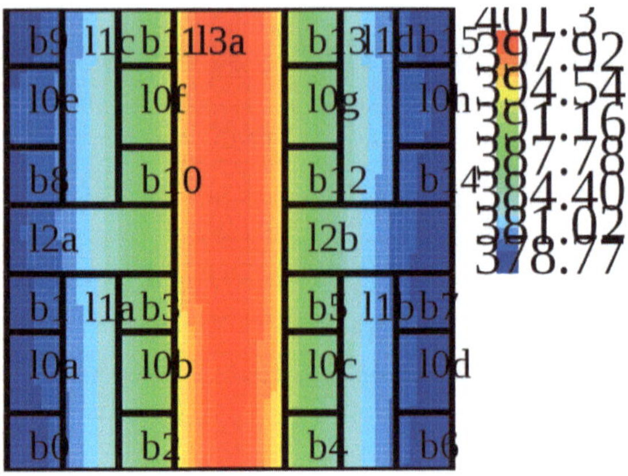

Fig. 7.6 Thermal profile of 3D Tree-based FPGA, tier 0 without using effective thermal conductivity

design and simulation 3D Tree-based FPGA. The floorplan (a) shows tier 1 design with clusters placed along with local interconnects. The thermal profile presented in Fig. 7.10 shows, with the help 3D thermal net using TTSVs and horizontal wires, the heat generated due the activity of tree level 3 and 6 interconnects switches has been transferred to other coldspots and to the heat spreader. The TSV distribution used to interconnect tier 0 and 1 are also shown in Fig. 7.10. The high temperature spots shown in Fig. 7.10 is relative. The heat transfer take place through metal wires and TTSVs placed between tier 0 and 1.

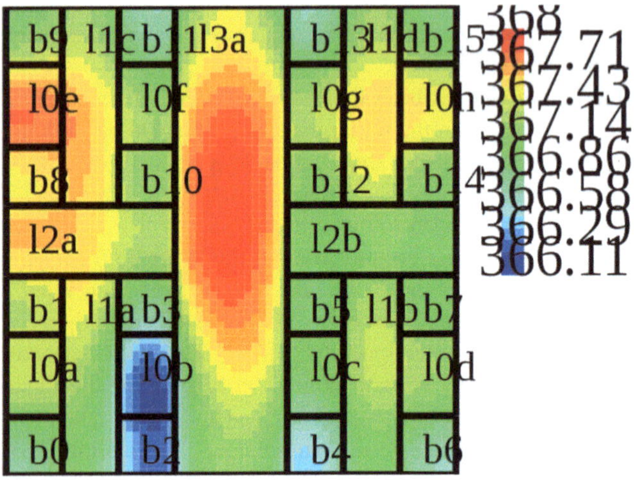

Fig. 7.7 Thermal profile of 3D Tree-based FPGA, tier 0 with effective thermal conductivity

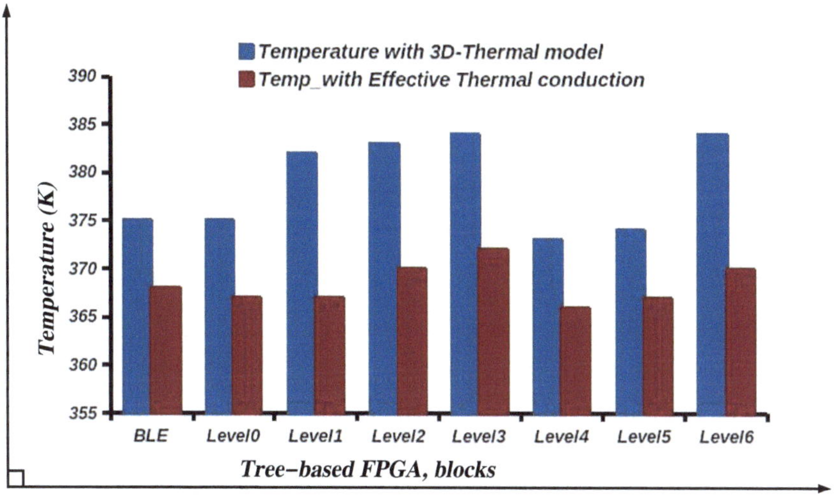

Fig. 7.8 Temperature values of functional units with and without effective thermal conductivity

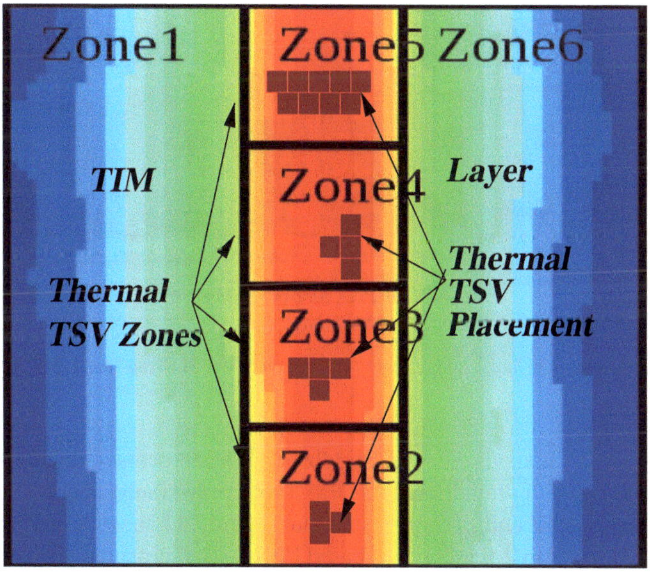

Fig. 7.9 Thermal Interface layer (TIM) with thermal TSV placement zones

The estimated values of inter-layer temperature is optimized by considering area and spatial distribution of TSVs and Hotspot zones. The TSVs located at theHotspot zones are effectively used as a 3D thermal net with help of local vias in the metal layers to transfer heat from tier 0 to tier 1 layer and also within the layer. The 3D thermal model considers the impact of via fill material based the type of technology used to

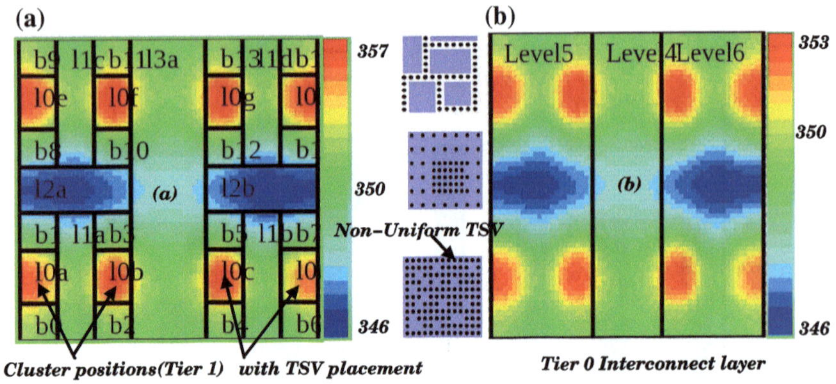

Fig. 7.10 Two-tier floorplan with thermal profile of 3D Tree-based FPGA along TSV distribution

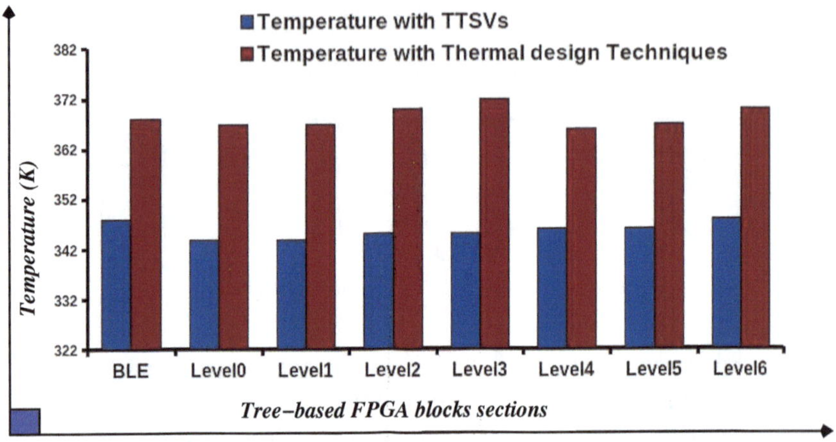

Fig. 7.11 Measured inter-layer temperature results from 2 tier 3D Tree-based FPGA

manufacture TSVs, like via-first, via-middle or via-last process. The via-first process use tungsten, while via-middle process use doped poly-silicon and via-last process use copper for via fill and SiO_2 for isolation. Figure 7.11 shows the temperature at different Tree levels in 2-tier 3D Tree-based FPGA. The measured peak temperature of 2D Tree-based FPGA is 73 °C and average temperature is 70 °C.

7.7 Summary

The main contribution of this work is the development of dedicated 3D thermal model for 3D Tree-based FPGA architectures. The model takes into account of in-homogeneous localized heating, heating exchange with the layer, heat transfer through external surface of the device, inter-tier heat transfer, dedicated hotspot zones assisted with help of thermal-TSVs and thermal design techniques. We have developed a 3D thermal CAD tools for the 3D FPGA model which is tested using two-tier 3D stacked Tree-based FPGA with variable TSV distribution, density, and materials. The 3D thermal model has been embedded into the thermal aware 3D-IC physical design tools to optimize thermal management at early stage of 3D FPGA design. The physical design and thermal simulation results demonstrated an accurate estimation of different temperature fields in each tier of the 3D FPGA chip. Furthermore the simulation results have shown the importance of using thermal design techniques and hardware based heat transfer in mitigating the heat produced within the 3D chip. Thermal design techniques are effective only when the number of devices in the 3D design are limited such as 3D ASIC, SoC, FPGA etc. However hardware based techniques are very effective when we have high number of high power devices such processors and memories integrated in 3D chip, for example MPSoC. In our work, we have integrated 3D thermal analysis tool to 3D physical design flow. This provides accurate estimation of temperature profile of each tier in a 3D stack.

References

1. J. Vlach, K. Singhal, in *Computer Methods for Circuit Analysis and Design* (Springer, 1983)
2. M.B. Kleiner, S.A. Kahn, P. Ramn, W. Weber, Thermal analysis of vertically integrated circuits, in *Proceedings of the IEEE International Electron Devices Meeting*, pp. 487–490 (1995)
3. T.A. Davis, I.S. Duff, An unsymmetric pattern multifunctional method for sparse LU factorization. SIAM J Matrix Anal. Appl. **18**(1), 140–158 (1997)
4. K.D. Hagen, *Heat Transfer with Applications*. (Prentice-Hall Inc, Upper Saddle River, 1999)
5. T.-Y. Chiang, S.J. Souri, C.O. Chui, K.C. Saraswat, Thermal analysis of heterogeneous 3D-ICs with various integration scenarios, in *Proceedings of the IEEE International Electron Devices Meeting*, pp. 681–684, Dec (2001)
6. K. Banerjee, S.K. Souri, P. Kapour, K.C. Saraswat, 3D-ICs: A novel chip design paradigm for improving deep-submicrometer interconnect performance and systems-on-chip integration, Proc. IEEE **89**, 602–633 (2001)
7. C.C. Liu, J. Zhang, A.K. Datta, S. Tiwari, Heating effects of clock drivers in bulk, SOI, and 3D CMOS. IEEE Trans. Electron Device Lett. **23**(12), 716–728 (2002)
8. B. Goplen, S. Sapatnekar, Efficient thermal placement of standard cells in 3D-ICs using a force directed approach, in *Proceedings of the IEEE/ACM International Conference onComputer-Aided Design*, Nov 2003, pp. 86–89
9. G. Chen, S. Sapatnekar, Partition-driven standard cell thermal placement, in *International Symposium on Physical Design*, CA, 2003
10. K. Skadron et al., Temperature aware microachitecture, in *International Symposium on Computer Architecture (ISCA)*, CA, 2003
11. S. Heo, K. Barr, K Asanovic, Reducing power density through activity migration, in *Proceedings of the ISPD*, pp. 217–222 (2003)
12. H. Su, F. Liu, A. Devga, E. Acar, S. Nassif, Full chip leakage estimation considering power supply and temperature variation, in *Proceeding of ISPD*, pp. 78–83, 2003
13. Altera Corporation, Thermal management for 90 nm FPGAs, Application Note. **358**, San Jose CA, 2004
14. J. Cong, J. Wie, Y. Zhang, A thermal-driven floorplanning algorithm for 3D-ICs, in *proceedings of the ICCAD*, pp. 306–313, 2004

15. Y. Han, I Koren, A. A. Moritz, Temperature aware floorplanning, in *2nd workshop on Temperature aware Computing systems (TACS-2)*, June 2005
16. K. Sankaranarayanan, S. Velusamy, M. Stan, K. Skadron, A case for thermal-aware floorplanning at the microarchitectural level. J. Instr.-Level Parallelism **7** (2005)
17. S. Gupta, M. Hilbert, S. Hong, R. Patti, *Techniques for producing 3D ICs with high-density interconnect* (Tezzaron Semiconductor, Naperville, IL, 2005)
18. S.R. Turns. Thermodynamics: concepts and applications, Cambridge University Press, New York (2006)
19. A. Telikepalli, Designing for power budgets and effective thermal management. Xcell J. **56**(56), 24–27 (2006)
20. M. Lin, A.E. Gamal, Y.-C. Lu, S. Wong, Performance benefits of monolithically stacked 3D FPGA, in *Proceedings of the 2006 ACM/SIGDA 14th International Symposium on Field Programmable Gate Arrays* (Monterey, CA, USA, 2006), Feb 22–24, pp. 113–122
21. P. Sundararajan, A. Gayasen, N. Vijaykrishnan, T. Tuan, Thermal characterization and optimization in platform FPGAs, in *International Conference on Computer-Aided Design (ICCAD-2006)*, pp. 443–447, 2006
22. F.P. Incropera, D.P. Dewitt, T.L. Bergman, A.S. Lavine, Fundamentals of heat and mass transfer, Wiley, New York, 2007
23. D.M. Jang, C. Ryu, K.Y. Lee, B.H. Cho, J. Kim, T.S. Oh, W.J. Lee, J. Yu, Development and evaluation of 3-D SiP with vertically interconnected through silicon vias (TSV), in *Proceedings 57th Electronic Components and Technology Conference, ECTC-07*, Reno, NV, pp 847–852, 2007
24. A. Jain, R. Jones, R. Chatterjee, S. Pozder, Z. Huang, Thermal modeling and design of 3D integrated circuits, in *Inter-society Conference on Thermal and Thermomechanical Phenomena in Electronic Systems*, pp. 1139–1145, 2008
25. A. Gayasen, V. Narayanan, M. Kandemir, A. Rahman, Designing a 3-D FPGA: switch box architecture and thermal issues. IEEE Trans. Very Large Scale Integr. VLSI Syst. 16(7), 882–893 (2008)
26. K. Siozios, A. Bartzas, D. Soudris, Architecture level exploration of alternative schmes targeting 3D FPGAs: A software supported methodology. Int. J. Reconfig. Comput. (2008)
27. J.L. Ayala, A. Sridhar, V. Pangracious, D. Atienza, Y. Leblebici, Through silicon via-based grid for thermal control in 3D chips, NanoNet, pp. 90–98 (2009)
28. A. Sridhar, A. Vincenzi, M. Ruggiero, T. Brunschwiler, D. Atienza 3D-ICE: Fast compact transient thermal modeling for 3D ICs with inter-tier liquid cooling, in *IEEE/ACM International Conference on Computer-Aided Design (ICCAD)*, pp. 463–470, 2010
29. J.-S. Yang, K. Athikulwongse, Y.-J. Lee, S.K. Lim, D. Pan, TSV stress aware timing analysis with applications to 3D-IC layout optimization, in *ACM Design Automation Conference* (2010)

Chapter 8
Physical Design and Implementation of 3D Tree-Based FPGAs

Abstract The semiconductor industries current enthusiasm for 3D-ICs is widespread and well warranted, but designing those 3D devices presents a challenge. Normal 2D design tools are thoroughly honed and refined over many years, nonetheless fail to address some of the critical issues of 3D-IC design. A new 3D-IC design process is evolving gradually from the 2D heritage. Today there are tools to handle a complete back-end flow and strides are being made to enable true 3D design and implementation using TSVs. In this chapter we discuss the design algorithms and techniques to develop 3D physical design tools and use of these tools to design and fabricate 3D stacked Tree-based FPGAs. This chapter starts with development of VHDL code generator and continue to the development 3D layouts of Tree-based FPGA using the 3D physical design tools developed for 3D FPGA design. A new CAD tool set for 3D physical design and verification based on Global Foundries 130 nm technology node modified to use Tezzaron's TSV technology is also developed and presented in this chapter. Through this chapter we addressed few specific issues 3D designers often encounter dealing with tools that are not specifically designed to meet their needs. We also presented few additional 3D design support tools such as 3D LVS/DRC to verify the LVS of the partitioned and merged 3D designs.

8.1 Introduction

The semiconductor industries's current enthusiasm for 3D-ICs is widespread and well warranted, but designing those 3D devices presents a challenge. Normal 2D design tools are thoroughly honed and refined over many years, nonetheless fail to address some of the critical issues of 3D-IC design. A new 3D-IC design process is evolving gradually from the 2D heritage. Tezzaron designed its first 3D integrated circuits in 2003, the designers used standard 2D CAD tools and cobbled together a 3D DRC/LVS flow based on scripts. Today there are tools to handle a complete back-end flow and strides are being made to enable true 3D design and implementation using TSVs. The major performance and power bottleneck of the FPGA design and manufacturing industry is the programmable interconnects and routing resources of

V. Pangracious et al., *Three-Dimensional Design Methodologies for Tree-based FPGA Architecture*, Lecture Notes in Electrical Engineering 350, DOI 10.1007/978-3-319-19174-4_8

FPGA, which have been found to account for up to 80 % of the total delay [1], up to 85 % of the total power consumption [2] and 90 % of the chip area when both local and global interconnects are considered [3, 4]. The 3D-IC physical design and manufacturing technology has emerged as one of the most promising solutions for overcoming the challenges faced by the interconnects and integration complexity in modern circuit designs [5], while the planar CMOS technology scaling is facing the eventual physical limits. TSVs as the key enabling technology element for 3D integration is currently being actively evaluated as a potential solution to reduce the interconnect delay and increase the logic density of FPGAs.

8.2 3D Tree-Based FPGA Design Requirements

Considering the area, delay and power consumption overhead, the programmable routing resources are the key design element in FPGA design. We selected Tree-based FPGA architecture using a Butterfly-Fat-Tree (BFT) based interconnect network for the target FPGA architecture due to two reasons: (1) feasibility of increasing logic density and (2) 56 % reduction of the total area compared to Mesh-based FPGA architecture [6]. The complexity and challenges associated with the 2D physical design and implementation of Tree-based FPGA architectures presented in [7, 8]. By selecting to work with Tree-based FPGA architecture and 3D technology, we strive for the feasibility of optimizing the interconnect area requirement and power consumption of 3D Tree-based FPGA designs by properly tailoring the structure and development strategy of partition and implementation of the Tree-based multilevel programmable interconnect using BFT based network topology. The 3D Tree-based interconnect design platform is an enabler for manufacturing fast and high density FPGA to meet the needs of modern circuit designs and prototyping.

8.2.1 Why Tree-Based Interconnect and Not Mesh

We have seen numerous studies [9–13] shows that the switch blocks (SBs) is the most area-consuming unit compared to other design elements in 2D Mesh-based FPGAs and this situation is becoming even more worse in 3D Mesh-based FPGAs because the TSVs are located on 3D-SBs. Although the design and manufacturing engineers are trying to reduce TSV dimensions, the minimum feature size on the die is also shrinking. Therefore, the TSVs are expected to remain larger than wire dimensions in metal layers within the die [14, 15]. Moreover it has been reported in [16] that the TSV utilization is actually quite low if the 3D-SBs are with full vertical connectivity in use. The experimental and simulation results presented in recent publications point out that the utilization of TSVs is actually very low in 3D Mesh-based FPGAs [16] with full vertical connectivity, which motivates us to explore new interconnect topology and architecture with better optimization flexibility to achieve

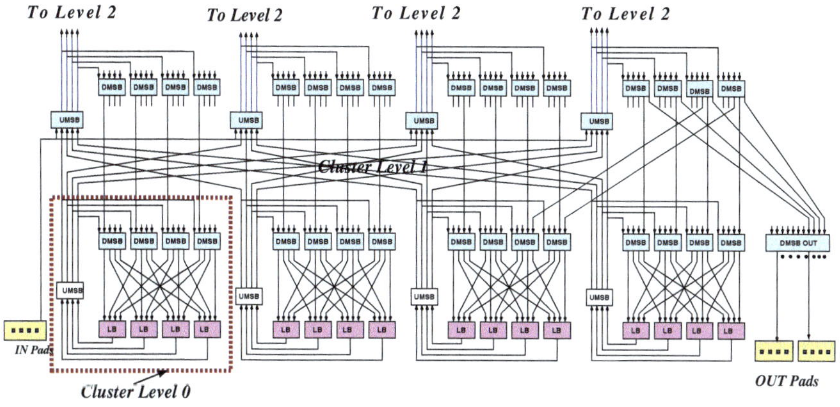

Fig. 8.1 An illustration of two-level Tree-based FPGA interconnect structure

higher speed, low power consumption, reduced silicon footprint and increased logic density to which the gap between FPGAs and ASICs can be minimized. Hence we selected Tree-based FPGA architecture with BFT based interconnect topology for 3D design and implementation. Our previous works [6, 17, 18], also confirms the multilevel BFT based 2D interconnect topology is able to reduce 59 % of the total switches and save 56 % of the total FPGA area compared to Mesh-based FPGA with identical logic density and array size [6]. Considering the challenges associated with 2D physical design of Tree-based FPGA [7, 8], we proposed two different network partitioning methodology to design and implement high density 3D FPGAs based on Tree-based interconnect network. We developed complete set of tools and technologies to conduct 3D design feasibility study and interconnect network characterization methodologies to build high performance 3D re-configurable systems based on Tree-based interconnect and a comparison procedure has been put in place to validate the advantages of 3D Tree-based FPGA over 3D Mesh-based FPGA architectures.

In a Tree-based FPGA architecture [6, 18], the programmable interconnects are arranged in a multilevel network with the switch blocks (SBs) placed at different tree levels and the Logic Blocks (LBs) are grouped into clusters as illustrated in Fig. 8.1. The modified Tree-based FPGA architecture unifies two unidirectional interconnect network. The downward interconnection network is based on a butterfly-fat-tree style interconnect topology with a linearly populated downward mini switch boxes (DMSBs) and unidirectional wires. Similarly the upward interconnect network uses upward mini switch boxes (UMSBs) to connect logic block outputs to all DMSBs and further to higher levels of the Tree. The number of DMSBs of a cluster located at level ℓ is equal to the number of inputs of a cluster located at level $\ell - 1$. The upward network also uses BFT topology to connects LBs outputs to the DMSBs at each level. As shown in Fig. 8.1, we use UMSBs (Upward MSBs) to allow LBs outputs to reach a large number of DMSBs and to reduce fanout on feedback lines. The number of UMSBs of a cluster located at level ℓ is equal to the number of outputs of a cluster located at level $\ell - 1$. UMSBs are organized in a way allowing

LBs belonging to the same *owner cluster* to reach exactly the same set of DMSBs at each level. Thus positions, inside the same cluster, are equivalent, and LBs can negotiate with their siblings the use of a larger number of DMSBs depending on their fanout. As illustrated in Fig. 8.1, input and output pads are grouped into specific clusters and are connected to UMSBs and DMSBs, respectively. Thus, input pads can reach all LBs of the architecture, and output pads can also be reached by all the from different paths. Using UMSBs and DMSBs greatly enhances routability, but it increases the interconnect switches number. However this increase in number switches is compensated by reducing in/out signals bandwidth of clusters at every level using Rent's Rule [19] based wire-length optimization tool. In fact, netlists implemented on FPGA architecture often communicate locally (intra-clusters) and this fact can be exploited to reduce the bandwidth of signals with inter-clusters communication. A good estimation of netlists communication locality is given by Rent's Rule [19]. Based on this estimation authors in [20] showed that most netlist Rent's parameters range between 0.5 and 0.65.

$$IO = c.m^{\ell.p} \tag{8.1}$$

The Rent's parameter [19] p defined for a Tree-based architecture shown in Eq. 8.1. The Tree level is represented as ℓ and m is the cluster arity, c is the number of in/out pins of an LB and IO is the number of in/out pins of a cluster located at level ℓ. Intuitively, p represents the locality in interconnect requirements. If most of the connections are routed locally and only a few of them communicate to the exterior of a local region, p will be small. In Tree-based architecture, both the upward and downward interconnects populations depend on this parameter. As shown in Fig. 8.2, we can depopulate the routing interconnect by reducing the number of inputs of each cluster of level 1 from 16 to 10 and outputs from 4 to 3 ($p = 0.73$). In this case, if we consider an architecture with 2 levels of hierarchy, we get a reduction in interconnect

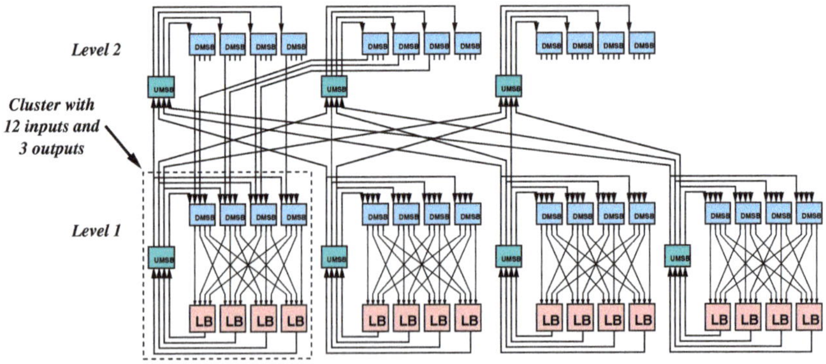

Fig. 8.2 An illustration of Tree-based interconnect optimization using Rent rule (Level ℓ with $p = 0.73$), based on Rent optimization, certain number inputs and outputs can be removed independently at each Tree levels

switches number from 521 to 368 (28 %). However, a reduction in the value of p reduces the routability of the architecture too. Thus we must find the best tradeoff between interconnect population and logic blocks occupancy. As shown in [21], the best way to improve circuit density is to balance logic blocks and interconnect utilization. In the case of Tree-based FPGA architecture, interconnect occupancy is controlled by p and logic occupancy factor is controlled by N, where N is the total number of LBs in the Tree.

8.2.2 3D Tree-Based Interconnect: A Requirement for High Logic Density

With the advent of sub-100 nm CMOS technologies, the design and prototyping cost of cell-based implementation have become exorbitant for most ASICs, making FPGAs increasingly popular. Current FPGAs, however, cannot meet the performance and power requirements of many ASICs due to their high programming overhead. As discussed in [1, 3, 4, 6] as much as 90 % of the FPGA area is occupied by the programmable routing resources. In addition to consuming most of the die area, the programmable interconnect also contributes significantly to total path delay in FPGAs [2, 22]. Furthermore programmable interconnect also contributes to the high power consumption of FPGAs, which has become a significant bottleneck to their adoption in many applications. As a result the FPGA performance is significantly worse in terms of logic density, delay, and power consumption than custom cell-based ASIC implementations. Studies [1, 2, 8] have estimated FPGA to be more than ten times less efficient in logic density, three times larger in delay, and three times higher in total power consumption that ASIC implementations. Although CMOS Technology scaling has greatly improved the overall performance of FPGAs, the performance gap between them and ASIC [2] has remained very wide mainly because the FPGA programming overhead and huge interconnect requirements. In [23] it is argued that the performance gap between FPGA and ASIC is becoming even greater in sub-100 nm technologies due to increase in parasitic elements of long wire routing segments. While the rate of increase in FPGA logic density has tracked that of ASIC implementations, the system frequency has scaled at a lower pace, and power consumption has risen to unacceptable levels.

A conceptually appealing approach to reducing the performance gap between FPGA and ASIC is to use true 3-Dimensional (3D) integration [9, 11, 24, 25], which increases the number of active silicon layers and optimizes the interconnect network vertically. Used correctly, 3D integration provides improved bandwidth and reduced wire length. In the best scenario, if we ignore inter-layer vias, the average wire length is expected to drop by a factor of $(N_{layers})^{1/2}$ [26]. The wire resistance and capacitance would drop proportionately; that is power would drop by a factor of $(N_{layers})^{1/2}$ and wire (RC) delay would drop by a factor of (N_{layers}). Hence for interconnect dominated architectures such as FPGAs, we expect a significant reduction in chip delay and power consumption. We know from our previous works [6, 8], Tree-based FPGA architecture achieves asymptotically fewer switch than the Mesh-based FPGA. Our previous works [6, 17, 18], confirms the multilevel BFT based 2D interconnect topology is able to reduce 59 % of the switches number and save 56 % of the total FPGA area [6] compared to Mesh-based FPGA with identical logic density and array size. However the 2D layout experiments shows the wire-length increases logarithmically as the Tree grows to higher levels [8]. It is difficult task to construct the layout of Tree-based FPGA architecture, specifically to be efficient for VLSI technology. The wiring structure of Tree-based architecture is sufficiently regular to permit a layout in $\Theta(N)$ area, which includes the wire connection nodes at different levels and corresponding switches using $O(log(N))$ wiring layers. Nonetheless this assumption of fixed number of wiring layers independent of device capacity is not in line the advance CMOS technology. Moreover we know the wirelength increases logarithmically as the Tree grows to higher levels. This will dictate the wire width growth in 2D VLSI model to be $O(log^2(N))$ and the overall 2D layout area to grow as $O(Nlog^2(N))$. Our approach to limit this wire growth and to generate realistic physical design model to be compatible to the modern VLSI technology. We proposed two different way to partitioned the Tree-based interconnect architecture to design and implement multi-tier 3D Tree-based FPGAs. However, before we move to the network design partitioning, we will explain the challenges and issues associated with the development of physical design for 2D Tree-based FPGA architecture.

8.3 2D Physical Design of Tree-Based FPGA

Designing the Tree-based interconnect routing resources is a major challenge for Tree-based FPGA. In order to maintain the hierarchy of Tree-based FPGA, a special layout methodology is used. We propose two ways to organize the layouts of Tree-based FPGA. The proposed methodology for physical design of Tree-based FPGA architecture include complete set of tools starting from VHDL to generation of layout (GDSII). The HDL generator is designed to generate VHDL code based on a hierarchical design approach that partitions the design into smaller sections, implement them separately and assemble them together at the final design phase. The physical design experiments are performed based on the layout generated using Global Foundries (GF-130 nm) 130 nm technology node provided by Tezzaron [27] modified to use Tezzaron's 3D TSV technology. Mentor's circuit simulator *Eldo* is used to estimate the wire delay switches and interconnection networks at different Tree levels. In this section we describe the challenges faced by the physical design of 2D Tree-based FPGA interconnect by using two different 2D physical design implementation methodes.

8.3.1 *Method 1: Coalesce Scalable Tree-Based 2D Layout Design*

The coalesce scalable 2D physical design methodology is developed in order to preserve the hierarchy of Tree-based FPGA. The Logic Blocks (LBs) are placed regularly along with interconnect switches according to the predefined tree parameters, such as network topology, dimension of the network nodes and floorplan specification. Thus, the layout method is scalable and capture different design and architecture constraints. Figure 8.3 illustrates the 2D floorplan of 4 × 4 (2 levels with cluster size 4) Tree-based FPGA. In order to spread the congestion and wire density over the FPGA surface, the different interconnect Tree levels were inter-waved to build the 2D floorplan, which is topologically equivalent to Tree-based FPGA. The 2D floorplan was designed with regular structure based on tiles, similar to Mesh-based FPGA. Each tile contains one logic block (LB) and a set of switches of different levels of the Tree. In this way, the layout obtained is scalable. The IOs of a cluster can be varied by increasing the switch size of a tile. However this layout is not comparable to industrial Mesh-based FPGA in terms of speed and performance due to larger wire lengths at higher levels. The physical design experiments revealed the wiring

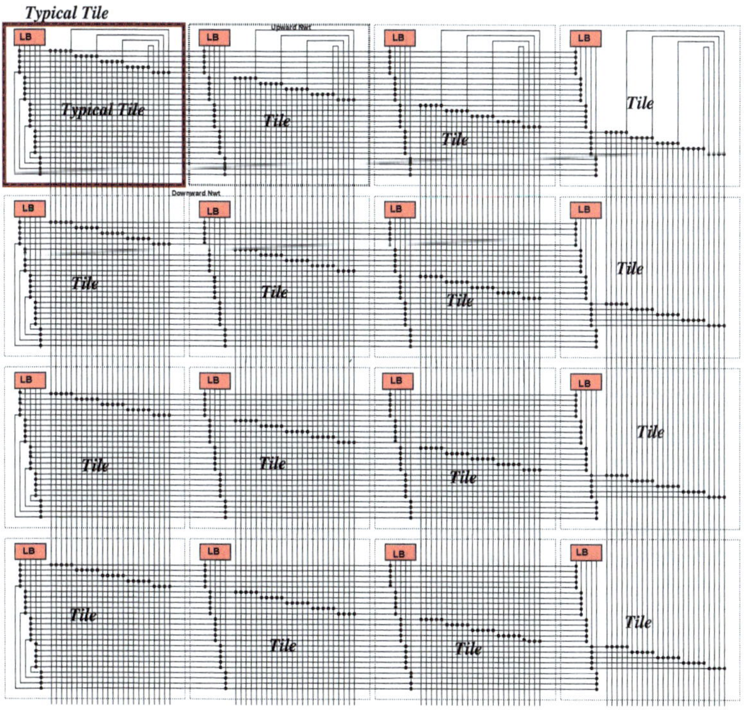

Fig. 8.3 2D optimized 4 × 4 Tree-based FPGA floorplan

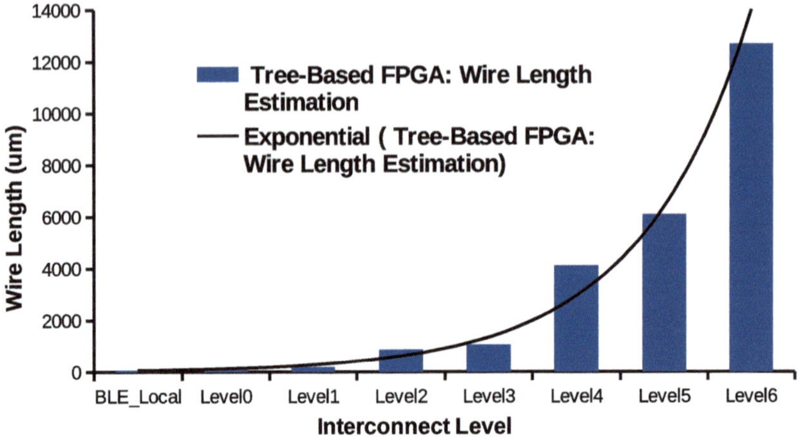

Fig. 8.4 Wire length extracted from Tree-based FPGA Layout

length increases exponentially as the Tree grows to higher levels. This is one of the major disadvantages of Tree-based FPGA architecture compared to Mesh-based FPGA, where the largest wiring distance is fixed. The wire length extracted from 2D Tree-based FPGA layout is illustrated in Fig. 8.4, which shows the exponential relationship of interconnect network of different Tree levels and wire length. The length of the local and global interconnect wires starting from *level 0 to 6* in μm, configured to Global Foundries (GF-130 nm) 130 nm technology, is presented in Fig. 8.4.

8.3.2 Method 2: Level-Wise 2D Tree Layout Design

To study and mitigate the long wire length issue of 2D layout at higher levels, a level-wise 2D Tree layout with 3D adaptability is designed. The 3D adaptability means the design can be partitioned horizontally or vertically based of the requirements to implement multi-tier 3D FPGAs. The disadvantage of level wise physical design is that, it is not scalable. The interconnect organization of the level-wise layout is arranged in such a way to bring together every cluster and its corresponding interconnect in order to form a hierarchical layout section with different tree level orientation to enable the feasibility study of multi-tier 3D Tree-based FPGA design. Figure 8.5 shows the floorplan of arity 4 Tree architecture with *level 0 and 1*. The LBs and interconnects are arranged in order to segregate the logic blocks and programmable interconnect of the Tree architecture into different tree level sections. Figure 8.6 illustrates the VLSI layout of *level 0 to 3* section of the Tree-based FPGA. This layout design offers the flexibility to partition the Tree interconnect at a certain level called the *break-point* based on optimization wire delay and transform the 2D layout into multi-tier stacked 3D FPGA chip. This will enable us to divide the

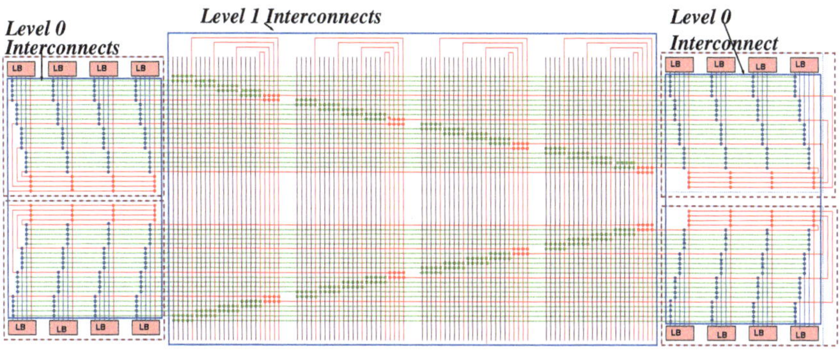

Fig. 8.5 Floorplan of Tree-based FPGA *levels 0 and 1*

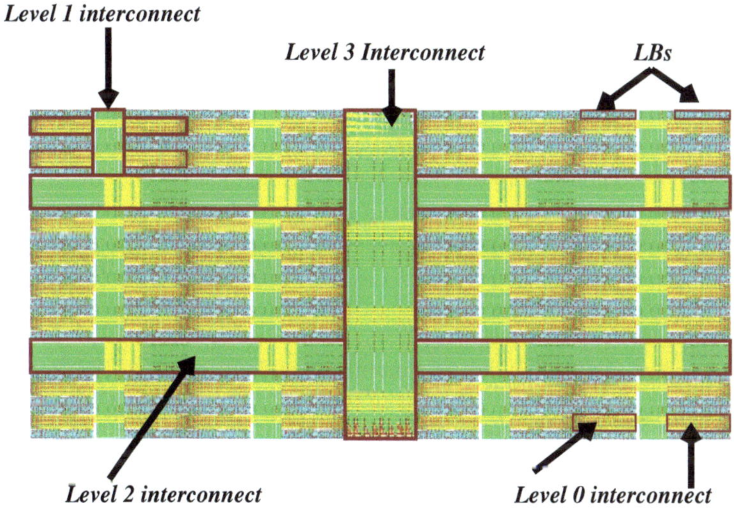

Fig. 8.6 Arity 4 Tree-based FPGA: VLSI layout of Tree *levels upto 3*

programming interconnects and place the major portion of SBs, I/0s and configuration memory above the *break-point* into a second active layer and to place logic blocks (LBs) and associated local programmable interconnects along with configuration memory into first active layer to implement two-tier 3D stacked Tree-based FPGA. Such an organization of FPGA primitives in a multi-tier chip design with LBs on one layer and SBs and I/Os belong to *break-point* and above on second layer will provide flexibility to improve the logic density and performance of Tree-based FPGA. However this is not possible with 2D tree-based layout illustrated Fig. 8.3, due to the tiled and rearranged Tree interconnection format. The long wire length and delay estimation of the Tree-architecture is done using the spice accurate circuit simulator *Eldo*, integrated along with 3D physical design flow.

8.4 Sub-path Timing Characterization

The subpath timing characterization is performed on 2D layouts generated using Global Foundries 130 nm Technology node. Wire lengths at different levels were evaluated from the layout and used Mentor's spice accurate circuit simulator *Eldo* to investigate interconnect delay. An accurate 130 nm transistor level technology models are used to investigate switch, interconnect delay and power estimation of Tree interconnect levels separately. The Tree-based FPGA architecture model used for timing characterization is illustrated in Fig. 8.7. The experimental model shown in Fig. 8.7 has only three levels, however we used a seven level Tree-based FPGA architecture for delay and power estimation. The experimental timing characterization model for tree interconnection network consists of two unidirectional the upward and downward interconnection timing path including an I/O interconnection is highlighted in Fig. 8.7. A sub-path connects a source to a sink and crosses several MSBs (Mini Switch Blocks). The number of sub-paths in the architecture is limited and depends on the number of levels. Consequently, given an architecture with n levels, we can isolate the n different sub-paths (symmetric structure). In Fig. 8.7 we show the 4 isolated sub-paths of an architecture containing 3 levels. Each architecture is composed of combinational sub-paths that either start from a logic block (Combinational/Sequential) or from an input pad pi and end on a logic block (Combinational/Sequential) or an output pad po. To ensure proper circuit operation, we also must take register setup-times t_{set} and sequential propagation delays d_{seq} into account (Sometimes denoted as *Clock-to-Q* delays). Classification of sub-paths and resulting delays is given below:

1. Combinational logic block → Combinational logic block
$d(p) = d(switches)$
2. Combinational logic block → Output-pad
$d(p) = d(switches) + d(po)$
3. Input-pad → Combinational logic block
$d(p) = d(pi) + d(switches)$

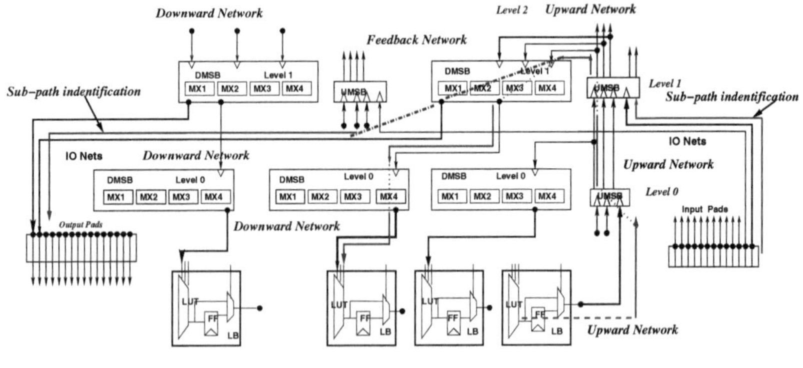

Fig. 8.7 Sub-path timing characterization setup shows two levels and IOs of Tree-based FPGA

4. Sequential logic block → Sequential logic block
$d(p) = d_{seq} + d(switches) + t_{set}$
5. Sequential logic block → Combinational logic block
$d(p) = d_{seq} + d(switches)$
6. Sequential logic block → Output-pad
$d(p) = d_{seq} + d(switches) + d(po)$
7. Input-pad → Sequential logic block
$d(p) = d(pi) + d(switches) + t_{set}$
8. Combinational logic block → Sequential logic block
$d(p) = d(switches) + t_{set}$

Delays on sub-paths depend on the length of wires connecting MSB and logic blocks. These lengths are extracted from the routed MFPGA layout. Figure 8.4 shows the wirelength extracted from a symmetric layout of a 7-levels Tree-based FPGA architecture (16K LBs). The basic tiles of the structure are:

- The LB that contains one multiplexer 16:1, one Flip-Flop and a bypass 2:1 Multiplexer,
- The MSB that contains 4 buffered multiplexers,
- The configuration Memory blocks composed of 16 SRAM cells,
- The decoder for configuration memory addressing.

These basic tiles are duplicated at each level to construct the hierarchy recursively. We abut those tiles using a symmetric H planning technique. Figure 8.8 shows the

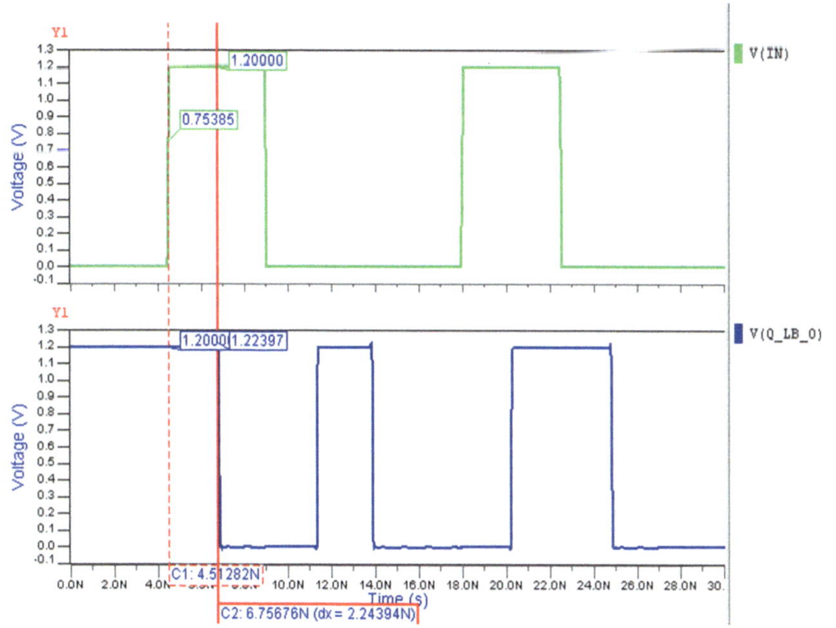

Fig. 8.8 Delay estimated using 2 Tree levels, combinational circuits without flip-flops

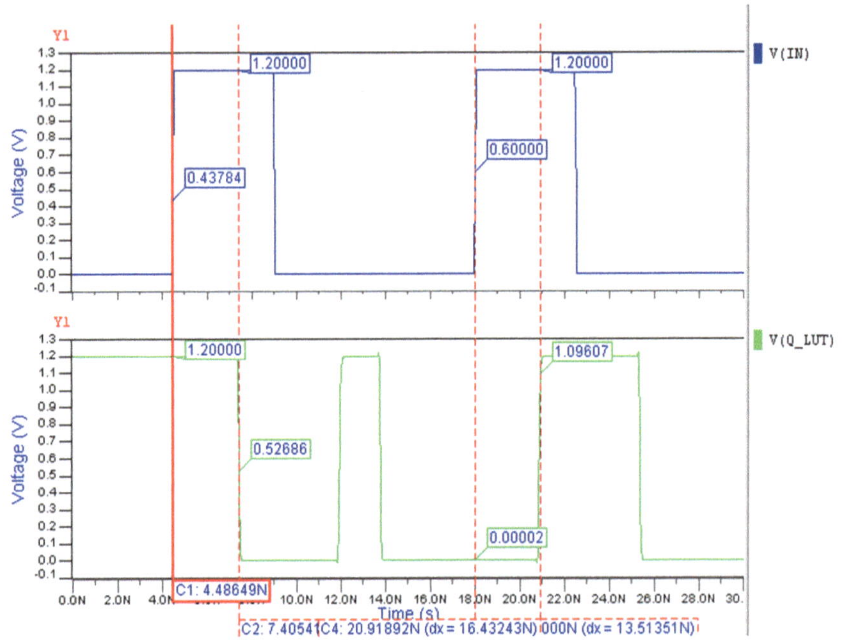

Fig. 8.9 Delay estimated using 5 tree Levels, Sequential circuits with flip-flops

eldo waveform for a 2 level Tree structure path including wire connections and switches. This shows an example of a sub-path which begins from an I/O pad and end at the out of a logic table. Figure 8.9 shows the delay extracted for 5 level Tree-based FPGA layout illustrated in Fig. 8.6. The sub-path used in this experiment begins from the output of multiplexers used inside logic blocks (LBs) which passes through the Flip-Flop and ends of the input of another logic block. All sub-path delay analysis is performed based on the sub-path classification mentioned above. The delay includes inter level wire connections, switches, buffers, SRAMs and logic blocks as mentioned above. Figure 8.10 shows the combined delay of wires and switches of upward programmable interconnection network at different levels of 7-level Tree-based FPGA. The sub-path delay analysis shows the logarithmic increase in path delay of Tree-based interconnect network. Figure 8.11 shows the network delay of feedback network at different tree levels. The feedback networks takes the output from the logic blocks and connected to next level of DMSBs or to the level as so by using UMSB network using BFT based network topology.

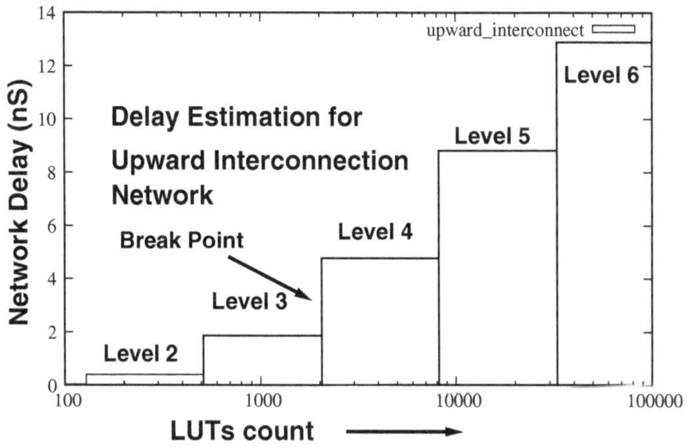

Fig. 8.10 Upward interconnect network delay estimation of 7 level Tree-based FPGA architecture

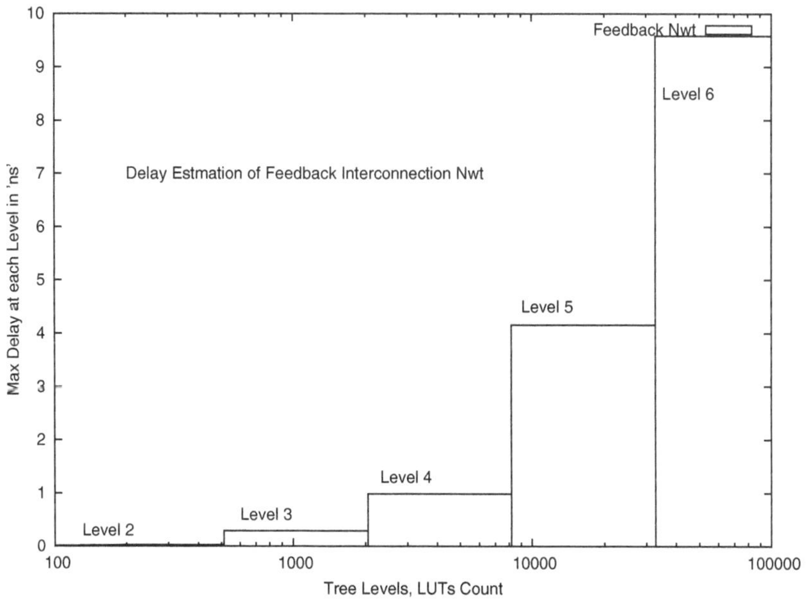

Fig. 8.11 Feedback network delay estimation of 7 level Tree-based FPGA architecture

8.5 3D Design Methodologies

We defined two types of network partitioning methodologies for the design and implementation 3D Tree-based FPGA: (1) *vertical partitioning:* the programmable interconnect network is partitioned vertically by placing the break-point at the highest

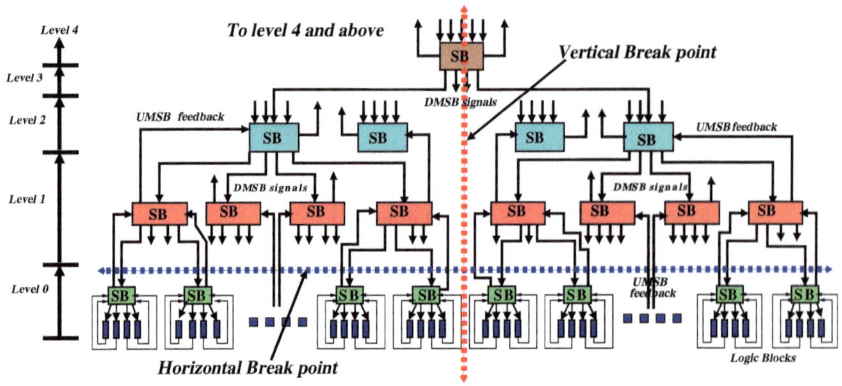

Fig. 8.12 Representation of programmable interconnect network partitioning (break point) of a four-level Tree-based homogeneous FPGA. Horizontal break-point: *blue dotted line*, Vertical break-point: *red dotted line*

level ℓ_v of the Tree-based programmable interconnect network to balance the silicon area and power consumption across multiple tiers of the 3D chip and (2) *horizontal partitioning:* the main objective is to optimize the critical path delay and improve logic density. The horizontal break-point is placed at a particular tree level ℓ_h based on the design and manufacturing constraints to achieve interconnect delay optimization using TSVs. The location of the level ℓ_v is always fixed at highest tree level, however the location of level ℓ_h is decided based on the architecture and wire delay requirements. Figure 8.12 illustrate both *vertical* and *horizontal break-points* in homogeneous Tree-based FPGA. As discussed already the main goal of horizontal partitioning is to limit the exponential increase in network delay as we increase the number of levels in the fat-Tree. In a horizontally-partitioned design the LBs and local programmable interconnects along with configuration memory is place in one layer and rest of the programmable interconnects and I/Os above the *break-point* along with configuration memory is placed in another layer. The advantage of such partitioning is that it provides flexibility in increasing logic density and optimizing network delay. However the horizontal partitioning methodology is not balanced in terms of area and power consumption. Previous work [3, 17] confirmed, that placing Hard-blocks like DSP slice and memory units at the higher level of Tree is more beneficial to optimize area and speed of heterogeneous FPGA. Figure 8.13 shows the heterogeneous version of Tree-based FPGA with the placement of hard-blocks and partitioning methods. In this case we utilized the white space left in active layer two to places hard-blocks to design heterogeneous Tree-based FPGA.

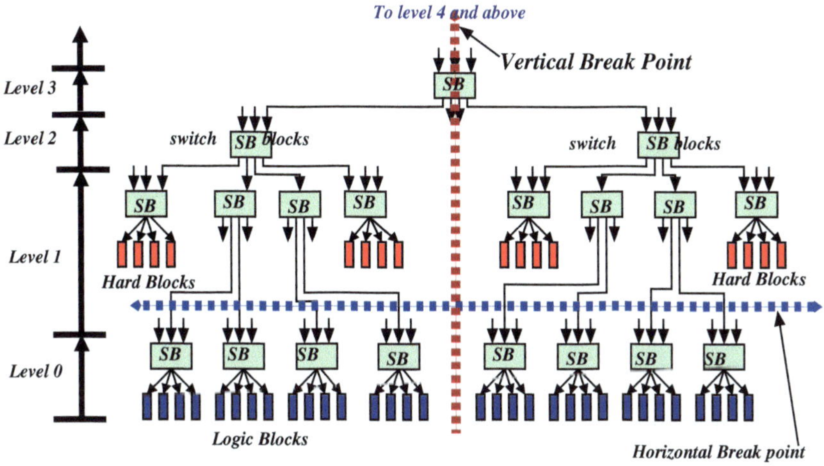

Fig. 8.13 Representation of programmable interconnect network partitioning (break point) of a four-level Tree-based heterogeneous FPGA. Horizontal break-point: *blue dotted line*, Vertical break-point: *red dotted line*

8.5.1 Vertical Partitioning

The main focus of vertical partitioning method is to balance the total silicon area and power consumption of the two-tier 3D homo/heterogeneous Tree-based FPGA equally between the active layers of the 3D stacked chip. The total number of LBs plus HBs and SBs are equally partitioned into multiple stacked tiers. Figure 8.14 shows the layout formation of two-tier homogeneous Tree-based FPGA. The break-point is set at the highest level of the Tree and interconnected using TSVs as illustrated in Fig. 8.14. Since the break-point is set at highest Tree level, the total number of TSVs required for a vertically partitioned 3D test chip with seven-Tree levels and 16K LUTs is 40,960 for a fully connected (Rent = 1) two-tier Tree-based FPGA. However we used Rent's Rule [19] based wire-length optimization model to find optimum number of TSVs and routing resources required for the design. Using Rent's Rule based wire-length optimization model we removed 17,203 TSVs from vertically partitioned Tree and 36,864 TSVs from horizontally partitioned two-tier 3D test chips with minimal impact on speed. The interconnect delay increases exponentially as the Tree grows to higher levels, the longest wire in 3D Tree-based FPGA located at highest Tree level is replaced by TSV and limited wire length optimization is possible at other levels due the hierarchical nature of upward interconnection network. In a vertically partitioned Tree, the downward interconnection network is more localized inside the owner cluster, however BFT based upward hierarchical interconnect network connects feedback to all cluster inputs. In a vertically partitioned Tree-based interconnect, only 50 % of the upward interconnects are realized using TSVs. This makes the *vertically partitioned* 3D FPGA ≈3.3 times slower in speed compared

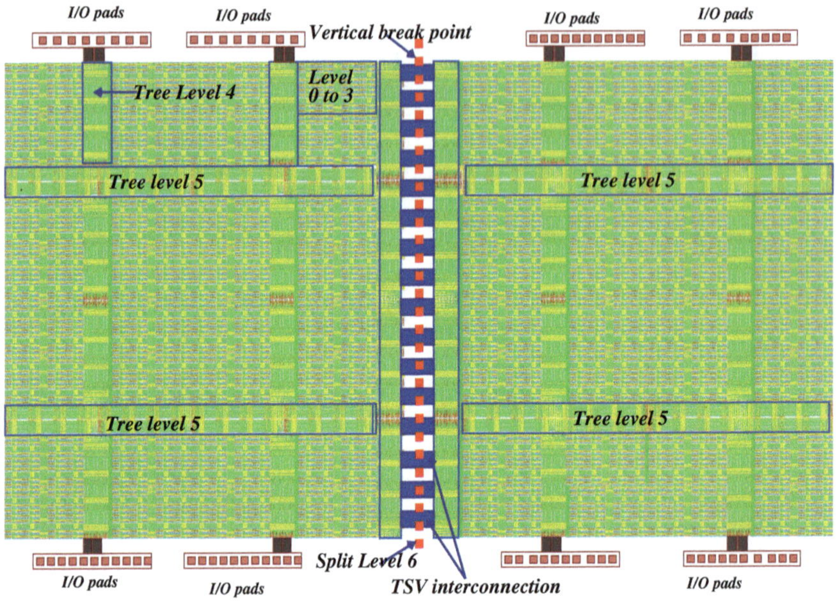

Fig. 8.14 Vertically partitioned two-tier homogeneous Tree-based FPGA with break point set at level 6. The *red dotted line* represent the break-point. HBs can be placed at any level of the Tree

to *horizontally partitioned* 3D Tree-based FPGA. Nevertheless, the advantages of vertical partitioning method compared to horizontal includes reduced chip area by 50 %, balanced power consumption across the tiers, reduced number of TSVs, and design complexity is minimized. Figure 8.15 shows the how the hard-blocks are connect to higher levels of the Tree and the impact of partitioning. Previous work shows [3, 17], it is better to stack hard-blocks higher level of Tree to optimize the area-delay product. In the case vertical partitioning, the hard-blocks are equally distributed on both tiers to balance the power consumption and area of the two-tier chip.

8.5.2 Horizontal Partitioning

In *horizontal* partitioning methodology, the location of the *break-point* is decided based on the estimated interconnect network delay. The interconnect delay of Tree-based architecture increases exponentially [6, 8, 28] as the Tree grows to higher levels. Horizontal partitioning methodology is introduced optimize the exponential increase Tree network delay. In horizontal partitioning methodology, the LBs and local interconnect levels below the *break-point* are placed in the bottom tier and programmable interconnect resources at levels above the break point and I/O pads are placed in the top tier of the 3D stacked chip as illustrated in Fig. 8.16. The Fig. 8.16

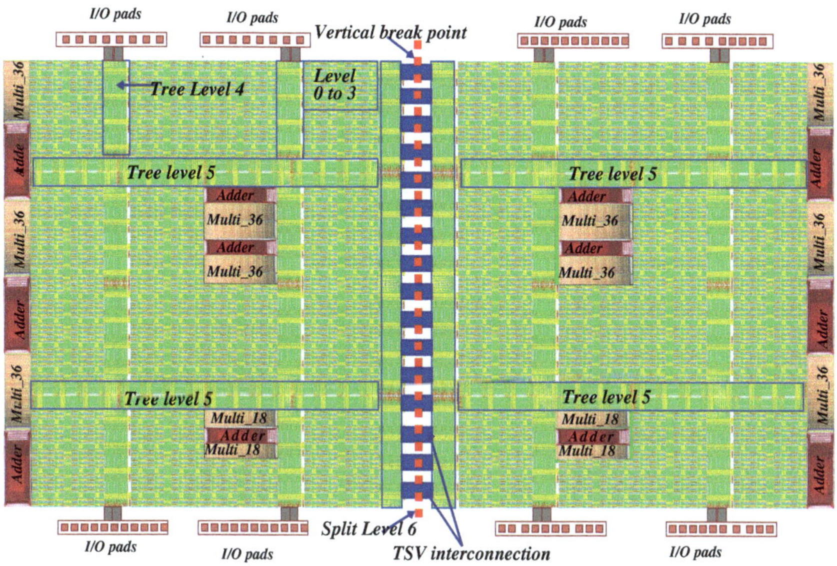

Fig. 8.15 Vertically partitioned 3D stacked heterogeneous Tree-based FPGA with break point set at level 6. The *red dotted line* represent the break-point. HBs can be placed at any level of the Tree

is an illustration of a two-level Tree architecture, in which the partition divides the LBs from the programmable routing resources. Figure 8.16 shows the 3D layout representation of two-tier homogeneous Tree-based FPGA, in which the I/Os are connected to Tree levels 4, 5 and 6 to reduce number of TSV used in the chip. The *horizontal break-point* (red dotted line) is set between level 3 and 4. The location of break point is decided based the delay measurements of tree levels discussed in Sect. 8.7. The delay measurements show the path delay of the programmable routing resources placed at the top tier is reduced 3 times and an overall path delay reduced 3.3 times compared to 2D layout. The interconnect path delay optimization is achieved by exploiting the design flexibility introduced by the segregation LBs and programmable routing resources into multiple tiers. Effectively almost 80 % of programmable routing network is placed on top of logic blocks is strongest point of *horizontal partitioning* methodology. The 3D test chip contains 16K LBs placed in tier 1 (bottom tier) with 65,536 vertical input pins and 16,384 vertical output (feedback) pins. For a fully connected 3D test chip requires 81,920 TSV communication between top (tier 0) and bottom tiers. Using Rent's Rule based wire-length optimization model, we removed 36,864 TSVs from horizontally partitioned two-tier 3D test chips with minimal impact on speed. We have approximately 20 % white space in tier 0 of horizontally partitioned 3D homogeneous Tree-based FPGA due to unbalanced hardware partition. As illustrated in Fig. 8.16, the partition is done in such a way to stack the programmable routing resource on top of logic-blocks and this method provides flexibility in increasing logic density and lowering critical path delay [25]. While designing 3D heterogeneous Tree-based FPGA, we used the white

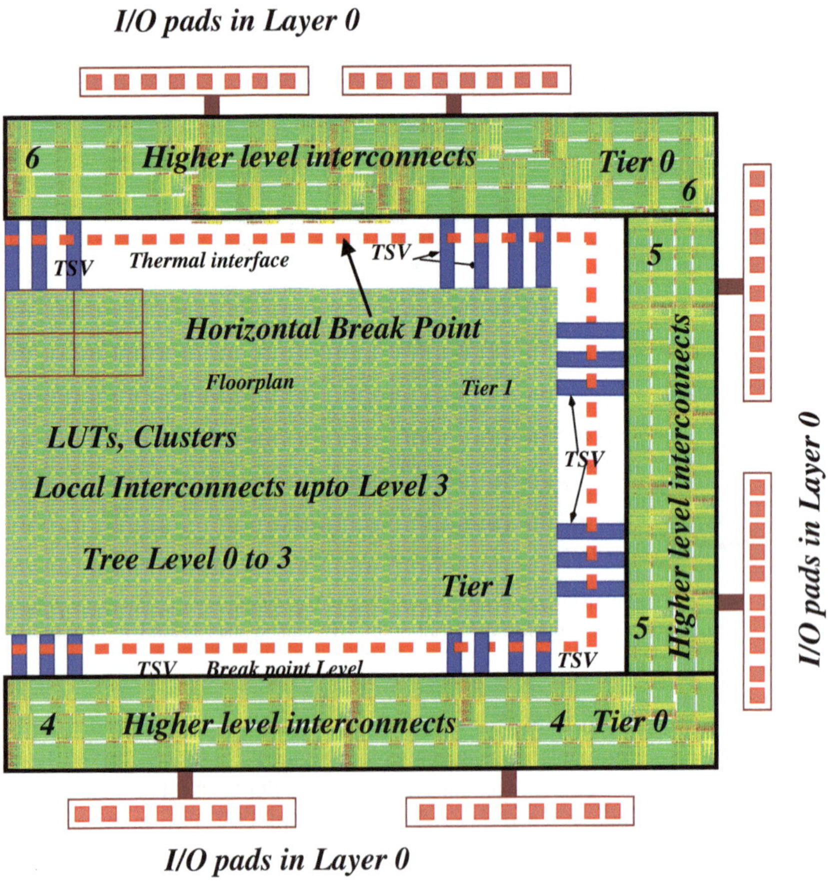

Fig. 8.16 Horizontal partitioning: break point between *level 3 and 4* of the 3D heterogeneous Tree-based FPGA. The *red dotted line* represent break-point. I/Os and HBs are placed in tier 0 along with high level interconnects

space available at tier 0 layer to stack *HBs* at multiple levels of Tree-based interconnect network as illustrated in Fig. 8.17. This is consistent with previous experiments conducted in [17] which leads to the conclusion that, placing *HBs* at higher levels of the Tree-based interconnect network leads to a better trade-off between area and speed of 3D heterogeneous Tree-based FPGAs.

8.5.3 Through Silicon via (TSV) Modeling

We used six metal 130 nm technology node provided by Global Foundries (GF-130 nm) that is modified to include TSVs according to the specification of *Tezzaron* Semiconductor. The GF-130 nm/Tezzaron process has one polysiliocn layer and 6

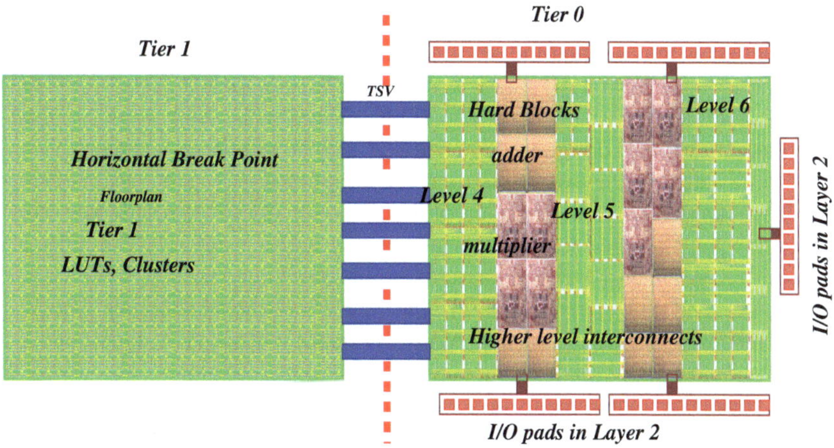

Fig. 8.17 Horizontal partitioning: break point between *level 3 and 4* of the 3D heterogeneous Tree-based FPGA. The *red dotted line* represent break-point. I/Os and HBs are placed in tier 0 along with high level interconnects

metal layers, however the last metal layer is reserved for wafer-to-wafer or TSV connections. Therefore, only five metal layers are available for design and all of them are thin metal layers intended for digital routing. Tezzaron's via-first 3D manufacturing process produces very small TSVs that are approximately 1.2 µm wide with 2.5 µm minimum pitch and 6 µm height [14]. The liner thickness is 100 nm and we used SiO_2 for liner deposition. The estimated values provided by Tezzaron for TSV resistance R_{TSV} and capacitance C_{TSV} are ≈600 mΩ and 15fF respectively. The wire delay estimation of tree levels for the 3D stacked Tree-based heterogeneous FPGA is extracted from the two-tier layout developed using Tezzaron Process and validated using Mentor's spice accurate circuit simulator *Eldo*. The break-point TSV delay is measured using the TSV model presented in [12, 29]. The TSV delay estimated using *eldo* is ≈28–32 pS. The wire delay estimation of tree levels for the 3D stacked Tree-based heterogeneous FPGA is extracted from the two-tier layout developed using Tezzaron Process and validated using Mentor's spice accurate circuit simulator *Eldo*. The break-point TSV delay is validated using *eldo* is ≈28–32 pS for Tezzaron TSVs. In tier 0, the spatial distribution of TSVs, SBs and HBs are rearranged in order to optimize the wire delay and temperature distribution at higher levels.

8.6 3D Tree-Based FPGA Physical Design Flow

Modern 3D designs, in which the active devices are placed in multiple layers using 3D integration technologies, are helping to extend the validity of technology scaling in today's nano era. However the progress in commercial 3D-ICs has been slow due to multiple reasons. One of them is the lack of appropriate physical design tools that takes into account the new constraints arise from the third dimension. In this section we present a 3D physical design tools flow for the design and implementation of Tree-based FPGAs. The 3D-IC design with TSVs do not require a revolutionary new 3D design system, however they do require new capabilities that need to be added to the existing toolsets for digital design, analog/custom design, and IC/package co-design. These capabilities should support the three key silicon realization goals:-unified design intent, abstraction, and convergence. The end goal is to optimize system cost with the shortest possible turnaround time. If 3D-ICs cannot be both cost and time effective, they will not enjoy widespread adoption. Above all, a comprehensive solution is needed. Many 3D stacks will combine digital and analog/RF circuitry, requiring a strong analog/mixed-signal capability. Because of the unique packaging requirements of stacked die, an IC/package co-design capability is a must. Additionally, fitting 3D-ICs on a board is challenging, requiring a capable PCB layout system with appropriate analysis tools. Thus, anyone who presents a complete *solution* must provide expertise in digital, analog, IC, package, and PCB design. 3D-IC design is a shared effort. The package designer knows where to put pins, but knows little about the design of the IC. The IC designer can place TSVs inside the die, but has limited knowledge of the package. The PCB designer will have to integrate the 3D IC package with other components on the board. Thus the design and development of 3D-ICs will require close collaboration and co-design among groups that have historically worked separately.

To design and develop 3D multi-tier Tree-based FPGA, we used Global Foundries 130 nm technology node modified to use *Tezzaron's* TSV technology [14, 27]. Figure 8.18 shows the overall 3D FPGA physical design flow to design and implement multi-tier 3D Tree-based FPGA. The design flow covers all areas of 3D design, including the design partitioning, merging multiple tiers (gds files) and design sign-off analysis. In addition, we also address few specific issues that 3D designers will encounter dealing with tools that are not specifically developed to meet their needs. The physical design process starts with the RTL description of Tree-based FPGA generated using VHDL code generator, which is developed to generate VHDL code based on a hierarchical design approach that partitions the design into smaller sections, which implement clusters separately and assemble them together at the final design phase. The physical design studies are performed using the layout generated with help of Global Foundries 130 nm technology node (Tezzaron 3D Design platform). Mentor's circuit simulator *Eldo* is used to characterize the wire delay and power consumption of switches and interconnection networks at different tree levels of 2D layouts. For the horizontally partitioned design, tier 1 contains LUTs and local programmable interconnects from Tree levels 0 to 3 (design2) and tier 0 contains programmable SBs and interconnects above the *break-point* along with IOs and HBs (design1) as illustrated in Figs. 8.16 and 8.17. We then used cadence design compiler to compile VHDL into structural Verilog for each die. The compiled Verilog is then input into Cadence Encounter to perform semi-automated physical design steps. This design tool augmented to test different 3D stacking methodologies. We used both Face-to-Face (F2F) and Face-2-Back (F2B) stacking methodology provided by

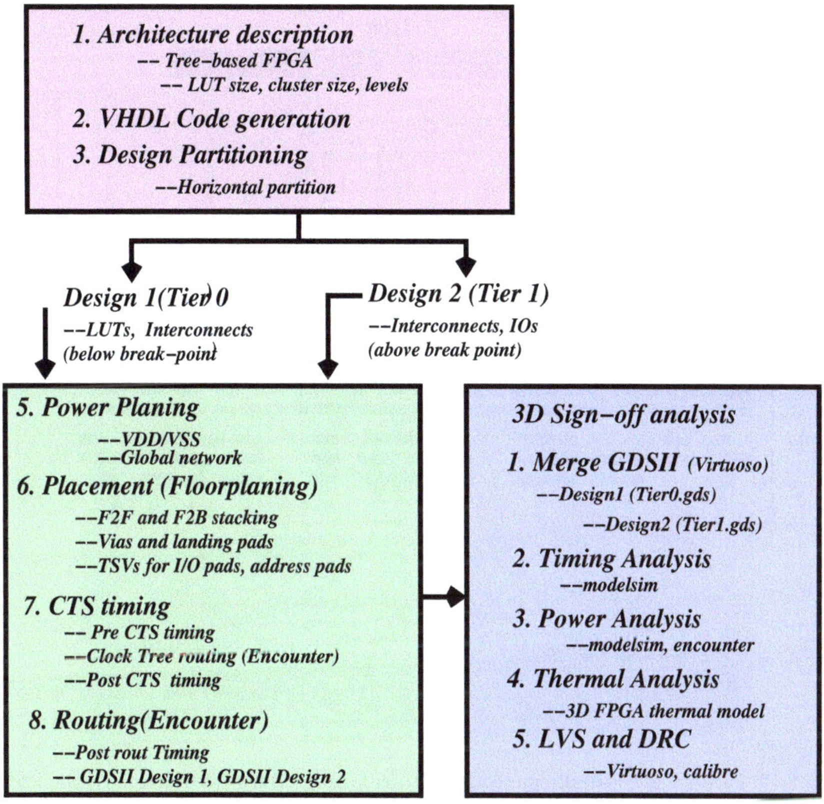

Fig. 8.18 3D FPGA physical design flow for design and analysis of horizontal and vertically partitioned 3D Tree-based FPGA using 2D design tools with additional add-on tools

Tezzaron's 3D design platform using via first TSV process. The Tezzaron's 3D stacking kit support two types of TSV structures: Super-Contact and Super-Via [14]. In our design we used Super-Contact, using tungsten via fill.

8.6.1 3D Stacking Methodologies

To experiment both F2F and F2B stacking methodologies, we conducted two experimental designs using two-tier stacked design, since we have only two layers in our 3D FPGA test chip. As illustrated in Fig. 8.19 in F2F stacking, the second wafer/die is flipped from left-to-right and bonded via copper thermal bonding using 3.4 μm Metal 6 pads with 5 μm pitch of the first wafer/die. After bonding the second wafer/die is

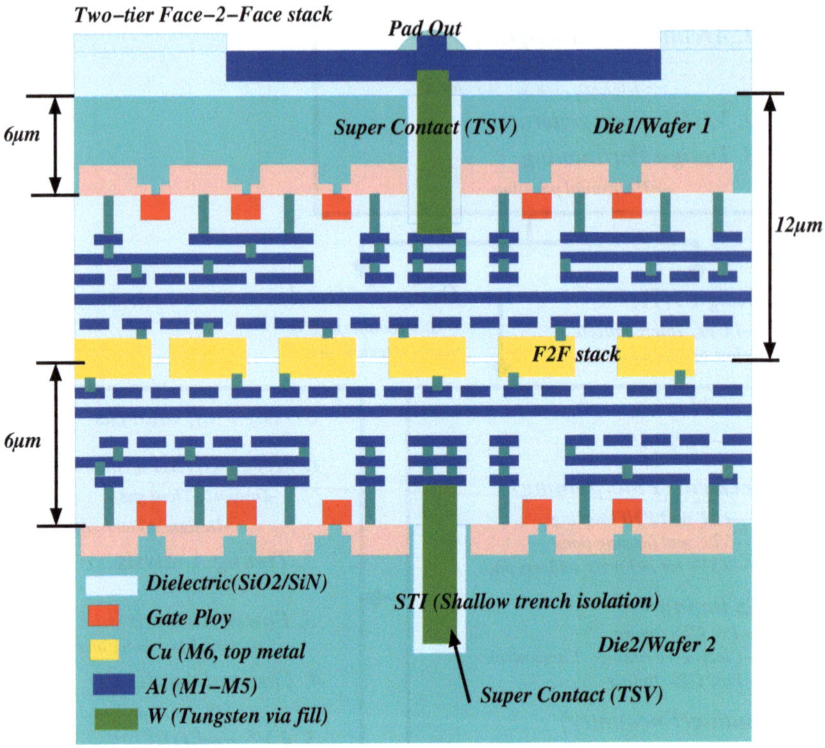

Fig. 8.19 Face-2-Face 3D stack representation of Tier 1 (design 2) and Tier 0 (design 1) of horizontal partitioned 3D Tree-based FPGA

thinned down to the TSV, that is 6 μm. After thinning, the second wafer/die is about 12 μm thick, with ≈6 μm metal and wiring plus 6 μm of bulk silicon with TSVs, where the original thickness of substrate is more 700 μm. The insulation material between TSV and silicon is oxide with 1000 Å thickness. The I/O signals are routed through TSVs to the back surface of tier 0 and from there, they will be fanned out past the edge of the device to connect to I/O pads on the surface of the 3D FPGA chip. In the case F2B stacking the second wafer/die or the tier 0 is thin down to TSV first and after thinning the tier 0 die is about 12 μm thick with ≈6 μm metal and wiring plus 6 μm of bulk silicon with TSVs. After this step the first wafer/die or tier 1 with logic density is flipped and stacked on top of the second wafer/die or tier 0 using the exposed TSVs and copper thermal bonding pads as illustrated in Fig. 8.20. After stacking the two-tier F2B test chip is flipped again to make I/O, power and ground connections on tier 1 die. The via-first TSVs between tier 0 and 1 have their landing pads on Metal 1 and Metal 6 as illustrated in Fig. 8.21. The connection between via-first TSVs are made using local interconnection and vias in between adjacent dies as illustrated in Fig. 8.20. In general, for F2B stacking, the TSV depth and wafer thickness may vary according to the 3D manufacturing process specifications. The

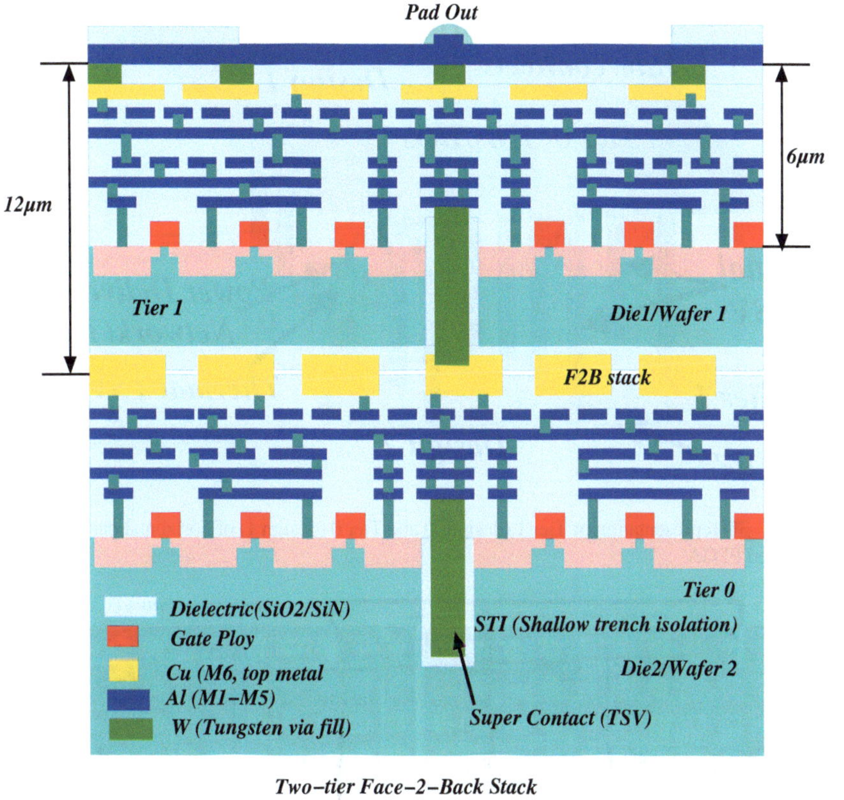

Fig. 8.20 Face-2-Back 3D representation of Tier 1 (design 2) and Tier 0 (design 1) of horizontal partitioned 3D Tree-based FPGA

copper signal and thermal landing pads includes *keep-out-zones* uniformly located around them to reduce coupling effects and CTE stress on active devices located around the TSVs, which is essential to maintain the 3D system performance.

8.6.2 3D FPGA Placement and Route

The place and route for two-tier 3D heterogeneous Tree-based FPGA is performed using Cadence Encounter. Figure 8.21 illustrates the pictorial representation of the logic units and interconnection switch blocks along with TSVs are arranged in a horizontally partitioned two-tier 3D stacked FPGA using F2B stacking configuration. The aim is to design and implement high density multi-tier 3D FPGA. To build such a multi-tier chip, we need to develop both F2F and F2B type of stacking as illustrated in Figs. 8.19 and 8.20. For the 3D FPGA demonstrator, we used a seven-level

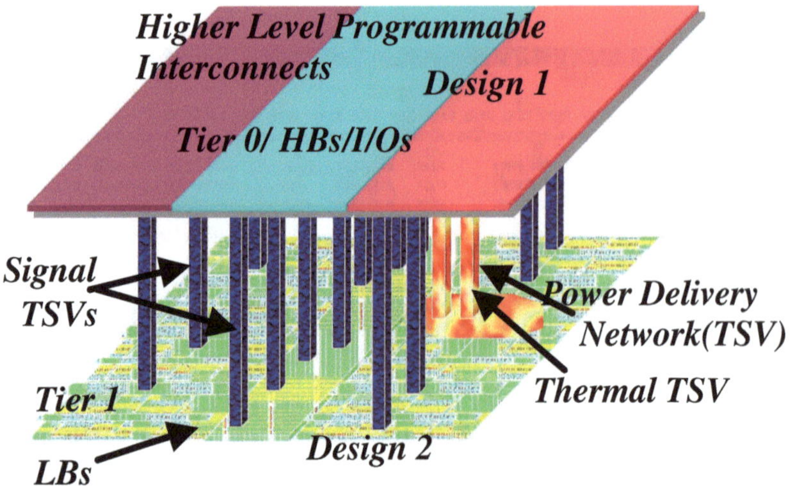

Fig. 8.21 3D representation of Tier 1 (design 2) and Tier 0 (design 1) of horizontal partitioned 3D Tree-based FPGA

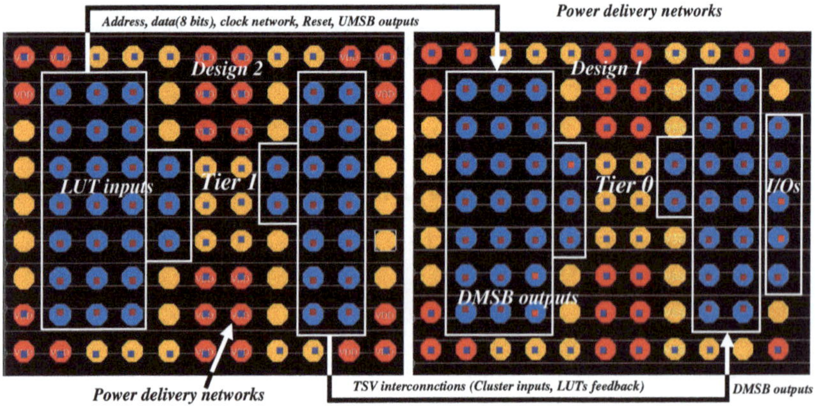

Fig. 8.22 TSV assignment on Tier 1 (design 2) and Tier 0 (design 1) of the 3D stacked Tree-based FPGA for F2B stacking configuration

homogeneous and heterogeneous Tree-based FPGA architectures with horizontal break-point ℓ_h placed between levels 3 and 4. In this experiment, we designed a two-tier F2B and F2F 3D Tree-based FPGA. Figure 8.22 shows the placement of vertical interconnection (TSVs) between tier 0 and 1, that is used to connect switch-block (DMSB outputs) from tier 0 to cluster inputs (LUT inputs) at tier 1 using butterfly-fat-tree network topology. Communication between LBs in tier 1 and higher level routing resources in tier 0 are established using TSVs for F2B configuration and for F2F stack we used copper thermal boning pads. Any net that connects to F2F/F2B via and thus interconnecting circuitry from tier 1 to tier 0, is defined as 3D net. The

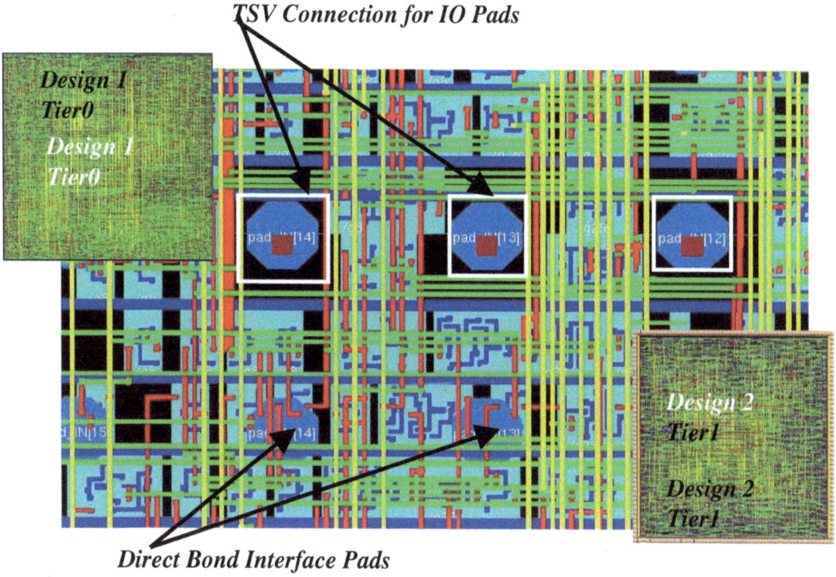

Fig. 8.23 TSV assignment on Tier 1 (design 2) shows the keep-out zone

individual designs for each die (tier 1 and tier 0) therefore must contain pins for all nets that cross the vertical interconnection boundary of the two-tier test chip with 7 Tree levels and 16K LBs and horizontal break-point placed between levels 3 and 4. The tier 1 (design2) contains only LBs and their communications to tier 0 (design 1). The interconnecting vias from tier 0 and 1 were manually placed on both dies for F2F using copper thermal bonding and TSVs are used only for off-chip I/O and power/ground connections.

One exclusive requirement that the Tezzaron TSV process imposes is on mandatory minimum TSV pitch of 250 μm throughout the entire wafer. This requirement forces us to include at least one TSV inside every 250 μm window in our design. According the Tezzaron 3D process, this is used for planarity of the wafer during Chemical and Mechanical Polishing (CMP) process. In F2F stacking, the die is thinned down to TSV after bonding without handling the wafer or die. Figure 8.23 shows the placement of TSV using metal 6 layers with *keep-out* zone in place. In certain locations, we manually inserted dummy TSVs before placement to meet this requirement. Figure 8.22 presents the TSV assignment of tier 0 and 1 dies of a cluster in the 3D stacked Tree-based FPGA test chip. In our horizontally partitioned 3D FPGA test chip, we have 7 Tree levels with arity set to 4 (cluster with 4 LUTs). In a fully connect $Rent = 1$ test chip contains 81,920 signal TSVs connections between tier 0 and 1 for fully connected FPGA chip. However we used interconnect optimization method using Rent's Rule to minimize the TSV count. The power and ground distribution networks (PDNs) are mainly generated using the strip and ring generation commands provided in cadence Encounter. Figure 8.24 shows the

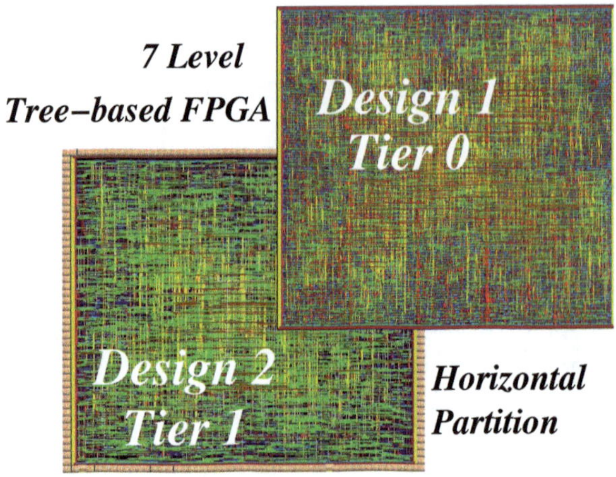

Fig. 8.24 Tier 1 (design 2) and Tier 0 (design 1) of the horizontally partitioned 7 level Tree-based FPGA with 16K LUTs

final singed-off two-tier layout design of horizontally partitioned seven-levels 3D heterogeneous Tree-based FPGA with 16K LUTs and 384 HBs connected at level 4 (ℓ_{HB}).

8.6.3 3D Design Sign Off Analysis

The verification of 3D-IC using TSVs requires a design flow for the design rule check (DRC) and Layout-Versus-Schematic (LVS), which is different from conventional planer CMOS process. In the case of DRC, it is not significant change, because in our 3D design flow, the two-tier 3D designs are treated as two independent planar dies, and each planar die could be checked separately using conventional design rule except for the TSV related design rule. The design rules for TSVs are provided by Tezzaron and they are fully compatible with conventional CAD tools such as Calibre DRC. At the present stage of technology development and 3D-IC applications, it is assumed that there are no interactions between the TSVs, and a model for a single TSV is provided by the foundries. For verification purposes, TSVs are treated as an LVS device (GDS based flow) or as a Via (LEF/DEF based flow), and the provided TSV model is used for downstream simulation. This assumption, however, may not hold true as the technology advances and the TSV densities and frequencies become high, accurate modeling and extraction of the interactions required. This new simulation model would require more accurate process description, accurate frequency dependent modeling and appropriate flow development to take advantage of the modeling accuracy. The DRC module used LEF/DEF based flow, in which the TSV is treated as via. The calibration with TSV profile, generate a simple R-C model for TSV and this model is used in simulation, if high frequency is not required. However the tools used for DRC did not have high frequency models. In normal case the high frequency model should be used in place of the R-C model generated.

Unlike DRC, LVS cannot be performed with an individual layer-based method. Since all tiers are electrically connected together, the schematic of the whole system should be compared across all layers of the system at the same time. However, the conventional CAD tools are unable to support 3D LVS, we developed pseudo 3D LVS method using conventional CAD tools, such as Calibre LVS. The most intuitive way is to add labels on the vertical connections. In the Tezzaron 3D-IC design process, every die has a pattern of uniformly distributed micro metal points (Supercontact) on both sides of the top and bottom tiers for wafer bonding. When two layers are stacked, all the metal points on one tier are connected to the matching metal points on the other layer. All vertical connection are made through this metal points and TSVs should be positioned on the metal points. By adding labels on all metal points of one tier and adding same labels on all metal points of the other tier, 3D LVS can be performed on side-by-side layouts, since Calibre-LVS recognizes the two nets having same net name as one connected net in the layout. To speedup this method, we used C programs to merge multiple streams of GDSII files into one integrated GDSII file and compare with the top level schematic as illustrated in Fig. 8.25. After completing DRC/LVS for both design1 and design2, we used the flow proposed in Fig. 8.25 to integrate the design1(tier0) and design2 (tier1). When Calibre-LVS compares this merged GDS file (tier01.gds) with top level schematic, it requires the same number of rule deck files as tiers. The rule deck file for the tier 0 is original rule deck file of conventional planar process and it has information on electrical connections inside tier 0 and other rule deck files are modified versions based on the original rule deck file. They have the information on connections inside their corresponding layers as well as electrical connections to adjacent layers. This methodology can be extended to check LVS of multi-tier 3D ICs. For more than two GDSII files we also have to modify the merger program to perform hierarchical integration of multiple (GDSII) layers.

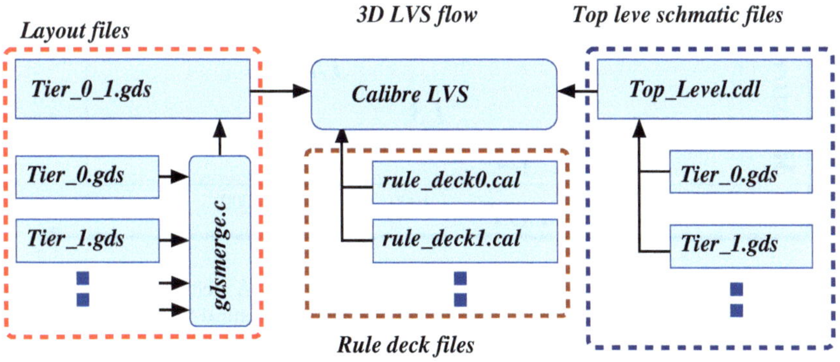

Fig. 8.25 Proposed LVS flow for 3D Tree-based FPGA

8.7 3D Timing Analysis

The timing analysis for *horizontal* and *vertical* partitioning is performed using the two-tier 3D Tree-based FPGA layout generated using Tezzaron 3D design platform. Figure 8.26 shows the timing analysis of seven level Tree-based FPGA architecture extrated from the 2D layout. As expected the interconnect delay increases exponentially as the tree grows to higher levels. Figures 8.27 and 8.28 shows the delay extracted from the 3D layout of the horizontally and vertically partitioned two-tier Tree-based FPGA respectively. We used *Eldo* to validate the delay values obtained from the layout experiments. The layout extraction for tier 0 and 1 provides the R and C values of inter-level interconnects and switches. Figure 8.27 shows the network delay of 3D horizontally partitioned layouts. The timing analysis horizontally partitioned layout shows the network delay reduced by 3.3 times compared to 2D layout as illustrated in Fig. 8.26. The *break-point* ℓ_h is placed between level 3 and 4. The delay between levels 3 and 4 is too small to be represented in Fig. 8.27, since TSV delay is ≈28 pS. Figure 8.28 shows the path delay estimated for vertically partitioned layouts. In this case the *break-point* ℓ_v is placed at level 6 and the other levels delay reduction almost negligible and due to this reason *vertically* partitioned Tree-based FPGA 3 times slower compared to *horizontally* partitioned Tree-based FPGA. The exploration setup shows *horizontal* and *vertical* partitioning methodology for two-tier 3D Tree-based FPGA performs 64 and 42 % better compared to 2D

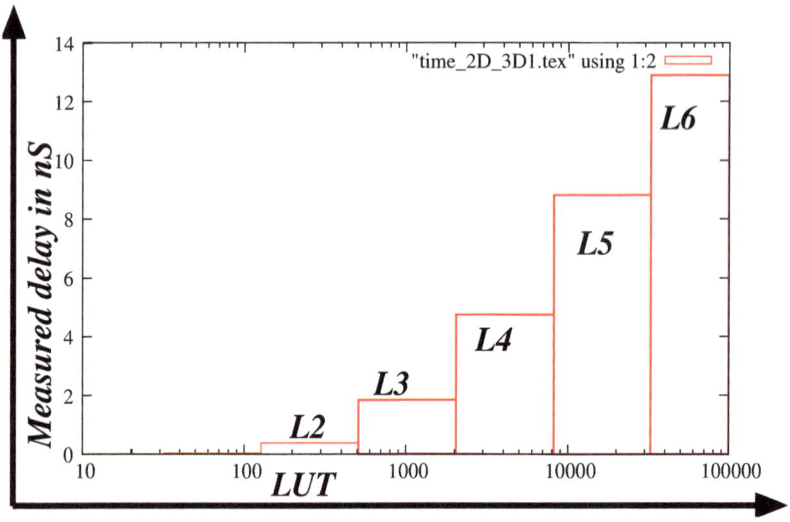

Fig. 8.26 Interconnect delay estimation of 7 level Tree-based FPGA architecture. The programmable interconnect is not pratitioned according to horizontal or vertical break-point 3D design

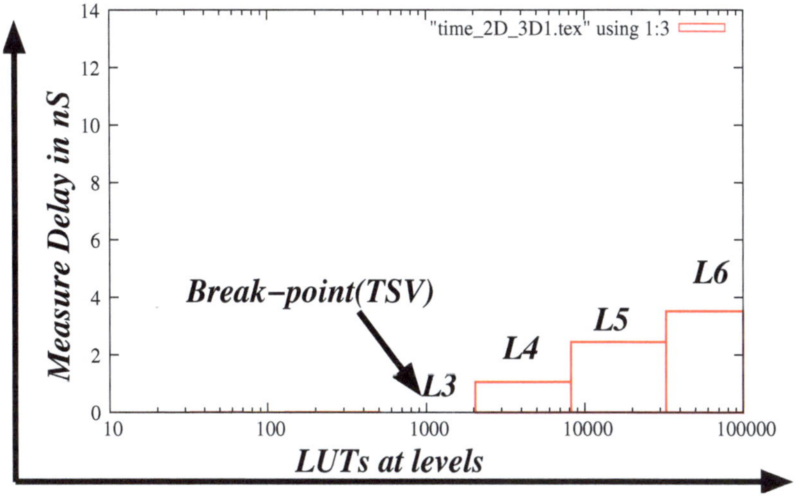

Fig. 8.27 Horizontal break-point interconnect delay estimation of 7 level Tree-based FPGA architecture. The programmable interconnect network is pratitioned horizontally between levels 3 and 4

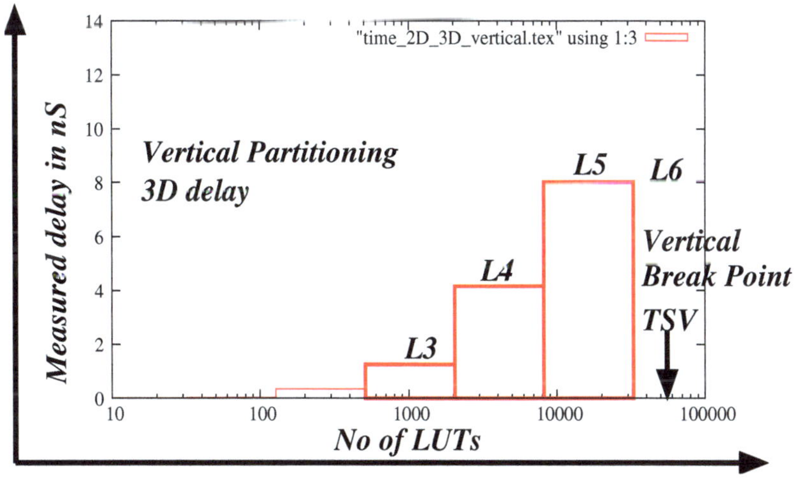

Fig. 8.28 Vertical break-point interconnect delay estimation of 7 level Tree-based FPGA architecture. The programmable interconnect network is pratitioned vertically and placed the break-point at the highest level of the tree

counterpart with identical logic density and size. As described in previous sections, the *horizontally* partitioned 3D Tree-based FPGA is design to optimize interconnect delay and the *vertically* partitioned 3D Tree-based FPGA is design to optimize area.

8.8 Summary

A 3D physical design and validation methodology for Tree-based FPGA architecture studied, implemented and presented in this chapter. Two different design partitioning methods namely *horizontal* and *vertical* also presented to support 3D design and implementation. The network partitioning is developed to optimize the area, power consumption and speed. A CAD tool set for 3D physical design and verification based on Global Foundries 130 nm technology node modified to use Tezzaron's TSV technology is also developed and presented. Through this chapter we addressed few specific issues 3D designers often encounter dealing with tools that are not specifically designed to meet their needs. We also presented few additional 3D design support tools such as 3D LVS/DRC, GDSmerge to verify the LVS of the partitioned and merged designs.

References

1. E. Ahmed, J. Rose, The effect of LUT and cluster size on deep-submicron FPGA performance and density. IEEE Trans. Very Large Scale Integr. (VLSI) Syst. **22**(3), 288–298 (2004)
2. F. Li, D. Chen, L. He, J. Cong, Architecture evaluation for power-efficient FPGAs, in *Proceedings of the ACM/SIGDA International Symposium on Field Programmable Gate Array*, Nov 2003, pp. 175–184
3. I. Koun, J. Rose, Measuring the gap between FPGAs and ASICs, in *FPGA'06*, Monterey, California, USA, 2006
4. S. Simon Wong, A. El-Gamal, The prospect of 3D-IC, in *IEEE Custom Integrated Circuit Conference (CICC)* (IEEE, San Jose, CA, 2009), pp. 445–448
5. A. Rahman, R. Reif, System level performance evaluation of three-dimensional integrated circuits. IEEE Trans. Very Large Scale (VLSI) Syst. **8**, 671–678 (2000)
6. Z. Marrakchi, H. Mrabet, U. Farooq, H. Mehrez, FPGA interconnect topologies exploration. Int. J. Reconfig. Comput. **2009** (2009)
7. A. DeHon, Unifying mesh- and tree-based programmable interconnect. IEEE Trans. Very Large Scale Integr. (VLSI) Syst. **12**(10), 1051–1065 (2004)
8. A. DeHon, R. Rubin, Design of FPGA interconnect for multilevel metallization. IEEE Trans. Very Large Scale Integr. (VLSI) Syst. **12**(10), 1038–1050 (2004)
9. C. Ababei, P. Maidee, K. Bazargan, Exploring potential benefits of 3D FPGA integration, in *Field Programmable Logic and Application*, vol. 3203 (Springer, Berlin, 2004), pp. 874–880
10. C. Ababei Y. Feng, B. Goplen, H. Mogal, T. Zhang, K. Bazargan, S. Sapatnekar, Placement and routing for 3D integrated circuits. IEEE Des. Test **22**(6), 520–531 (2005)
11. C. Ababei, H. Mogal, K. Bazargan, Three-dimensional place and route for FPGAs. IEEE Trans. Comput. Aided Des. Integr. Circuits Syst. **25**(6), 1132–1140 (2006)
12. K. Siozios, A. Bartzas, D. Soudris, Architecture level exploration of alternative schmes targeting 3D FPGAs: a software supported methodology. Int. J. Reconfig. Comput. **2008** (2008)
13. K. Siozios, V.F. Pavlidis, D. Soudris, A novel framework for exploring 3-D FPGAs with heterogeneous interconnect fabric. ACM Trans. Reconfig. Technol. Syst. **5**(1) (2012)
14. S. Gupta, M. Hilbert, S. Hong, R. Patti, *Techniques for Producing 3D ICs with High-Density Interconnect* (Tezzaron Semiconductor, Naperville, 2005)
15. A. Gayasen, V. Narayanan, M. Kandemir, A. Rahman, Designing a 3-D FPGA: switch box architecture and thermal issues. IEEE Trans. Very Large Scale Integr. (VLSI) Syst. **16**(7), 882–893 (2008)

16. C.-I. Chen, B.-C. Lee, J.-D. Huang, Architectural exploration of 3D FPGAs towards a better balance between area and delay, in *Design, Automation and Test in Europe Conference and Exhibition (DATE)* (IEEE, Grenoble, France, 2011)
17. U. Farooq, H. Parvez, Z. Marrakchi, H. Mehrez, A new heterogeneous tree-based application specific FPGA and its comparison with mesh-based application specific FPGA. Elsevier Microprocess. Microsyst. J. Appl. Reconfig. Comput. Spec. Issue **36**(8), 588–605 (2012)
18. V. Pangracious, Z. Marrakchi, E. Amouri, H. Meherez, Performance analysis and optimization of high density tree-based 3D multilevel FPGA, in *ARC11-2013, Reconfigurable Computing: Architectures, Tools and Applications Lecture Notes in Computer Science*, vol. 7806 (Springer, LA, USA, 2013), pp. 197–209
19. B. Landman, R. Russo, On a pin versus block relationship for partitions of logic graphs. IEEE Trans. Comput. **20**(12), 1469–1479 (1971)
20. J. Pistorius, M. Hutton, Placement rent exponent calculation methods, temporal behaviour and FPGA architecture evaluation, in *Proceedings of the International Workshop on System Level Interconnect Prediction*, Apr 2003, Monterey, California, USA, pp. 31–38
21. A. DeHon, Balancing interconnect and computation in a reconfigurable computing array (or, why you don't really want 100% LUT utilization), in *Proceedings of the 1999 ACM/SIGDA Seventh International Symposium on Field Programmable Gate Arrays*, 21–23 Feb 1999, Monterey, California, USA, pp. 69–78
22. D. Lewis et al., The stratix logic and routing architecture, in *International Symposium on Field Programmable Gate Arrays, FPGA-2003*, Feb 2003, pp. 12–20
23. P.S. Zuchowski, C.B. Reynolds, R.J. Grupp, S.G. Davis, B. Cremen, B. Troxel, A hybrid ASIC and FPGA architecture, in *Proceedings of IEEE/ACM International Conference on Computer Aided Design*, May 2002, pp. 187–194
24. K. Banerjee, S.K. Souri, P. Kapour, K.C. Saraswat, 3D-ICs: a novel chip design paradigm for improving deep-submicrometer interconnect performance and systems-on-chip integration. Proc. IEEE **89**, 602–633 (2001)
25. M. Lin, A. El Gamal, Y.-C. Lu, S. Wong, Performance benefits of monolithically stacked 3D FPGA, in *Proceedings of the 2006 ACM/SIGDA 14th International Symposium on Field Programmable Gate Arrays*, 22–24 Feb 2006, Monterey, California, USA, pp. 113–122
26. W.R. Davis et al., Demystifying 3D-ICs: the pros and cons of going vertical. IEEE Des. Test Comput. **22** (2005)
27. R. Patti, Advances in 3D memory and logic devices, in *IMAPS International Conference on Device Packaging, TA13*, Scottsdale, AZ, March 2010
28. Z. Marrakchi, H. Mrabet, C. Masson, H. Mehrez, Mesh of tree: unifying mesh and MFPGA for better device performances, in *NOCS*, pp. 243–252, 2007
29. V. Pavlidis, E.G. Friedman, Interconnect-based design methodologies for three-dimensional integrated circuits, in *Proceedings of the IEEE*, Jan 2009, pp. 123–140

Chapter 9
Three-Dimensional FPGAs: Future Lines of Research

Abstract In this book, a collection of 3D FPGA architecture exploration, physical design, implementation methodologies and CAD tools for Tree-based FPGA architecture has been described and evaluated. We developed new and practical 3D Tree-based interconnect topologies and architecture to improve the performance, area and logic density for modern FPGA designs. While the research work presented in this book is an important step in the development of 3D FPGAs and CAD tools that can support the design and implementation of modern FPGA architectures, there is much work that remains. Here we give few new research directions to improve the 3D FPGA technology described in this book to cover the gap between FPGA and ASIC and to make FPGA technology competent enough to reduce the exorbitant cost of design and manufacturing.

9.1 Introduction: 3D FPGA Research

In this book, a collection of 3D FPGA architecture exploration, physical design, implementation methodologies and CAD tools for Tree-based FPGA architecture has been described and evaluated. We developed new and practical 3D Tree-based interconnect topologies and architecture to improve the performance, area and logic density for modern FPGA designs. Mesh-based island style FPGA architecture is the most studied and researched architecture in academia and semiconductor industry and used their innovation and research efforts to improve the architecture performance in terms of area, speed and power consumption. Modern Mesh-based FPGA architectures have optimized interconnect topology, different segment lengths and well adapted algorithms to optimize circuit implementation. However, Tree-based FPGA architecture using a *fat-tree* network topology is a very old idea, which resembles a complete binary Tree, in which more than one interconnect exists between a node and its parent, resulting in a high efficient FPGA interconnect network. Despite its good properties in optimizing area and efficient utilization routing resources, Tree-based FPGA architectures are not been overlooked till now. Many research reports over time has provided evidence to suggest that Tree-based FPGA architecture requires less number of switches in circuit implementation due to its hierarchical interconnect

V. Pangracious et al., *Three-Dimensional Design Methodologies for Tree-based FPGA Architecture*, Lecture Notes in Electrical Engineering 350, DOI 10.1007/978-3-319-19174-4_9

network topology compared to island style Mesh-based industrial FPGA architecture. In our work, the comparison between Tree- and Mesh-based FPGA shows, Tree-based architecture saves 59 % of the total number switches and 56 % of the total area. High logic density is another advantage of Tree-based FPGA architecture. As logic density increase the number of levels in fat-tree also increase and this leads few major issues in maintaining the speed and area of chip. Previous research reports demonstrated the complexity of developing 2D physical design for N-node fat-tree based FPGA architecture. It was shown that a N-node fat-tree based FPGA architecture can be laid out in $\Theta(N)$ area using $\Theta(log(N))$ wiring layers. However this fact was established under the assumption of fixed number of wiring layers independent of device capacity and this does not match with the technology advances in modern VLSI technology. The second issue we faced with Tree-based FPGA architecture is the exorbitant interconnect delay, which increases exponentially as the size of the Tree increase because the inter-level wire-length increase as the tree grows to higher levels. Nevertheless this dissertation presents practical solution to those issues and provides precise and undisputed solutions to improve performance, logic density, area, power consumption and to reduce complexity in design, implementation and manufacturing of high performance FPGAs using 3D Tree-based interconnect network.

In this book we have studied the main advantages of three-dimensional (3D) integration technology, in which multiple tiers of active devices are stacked and interconnected using through silicon via (TSVs), such as short interconnects, high performance, high integration capacity and low power consumption. Our main issue with Tree-based interconnect is long wire-length at higher levels and to resolve this issue, we have proposed two main interconnect partitioning methods to design 3D *Butterfly-fat-tree* interconnect network, in which the long tree-based interconnect in partitioned into multiple tiers and stacked using 3D integration technology, in which the tiers are interconnected using TSVs. Though 3D VLSI technology is not very well advanced like 2D VLSI, a lot of research and demonstration from academia and semiconductor industry have been reported. The Chap. 2 discuss the current state-of-the art of 3D semiconductor technology and its capabilities.

9.2 Tree-Based Interconnect Partitioning

We propose two types of network partitioning methodologies to design and implement 3D Tree-based interconnect network topology.

9.2.1 Vertical Partitioning

The programmable interconnect network is partitioned vertically by placing the *break-point* at the highest level ℓ_v of the Tree-based programmable interconnect network to balance the silicon area and power consumption across multiple tiers of the 3D chip. The location of the level ℓ_v is always fixed at highest tree level since the tree is partitioned vertically. The vertical partitioning methodology is designed to balance the total silicon area and power consumption of the chip across the multiple tiers of 3D Tree-based FPGA. Another advantage of vertical partitioning is the number is TSV required is 50 % less compared to *Horizontal partitioning* method. The 3D FPGA demonstrator chip has 7 Tree levels and 16K LUTs require 40,960 TSVs for a fully connected case (Rent = 1) implemented using two-tier F2B 3D stacking methodology. We developed interconnect optimization models using Rent's Rule to find minimum numbers of TSVs and interconnects required and achieved 42 % reduction in total TSV count.

9.2.2 Horizontal Partitioning

The main objective is to optimize the critical path delay and improve logic density. The horizontal *break-point* is placed at a particular tree level ℓ_h based on the design and manufacturing constraints to achieve interconnect delay optimization using TSVs. The location of level ℓ_h is decided based on the architecture and wire delay requirements. In *horizontal* partitioning methodology, the location of the break point is decided based on the estimated interconnect network delay. The interconnect delay of Tree-based architecture increases exponentially as the Tree grows to higher levels. Horizontal partitioning methodology is introduced optimize the exponential increase Tree network delay. We set the *break-point* between level 3 and 4 and the network delay between level 4, 5, and 6 were minimized due to design flexibility introduced by partitioning the tree into two separate tiers of the 3D chip. The 3D test FPGA demonstrator chip has 7 Tree levels and 16K LUTs require 65,536 3D nets named as cluster inputs pins and 16,384 feedback networks pins. For a fully connected (Rent = 1) horizontally partitioned 3D homogeneous FPGA expected to have 81,920 3D nets required TSV communication excluding I/O pads. We developed interconnect optimization models using Rent's Rule to find minimum numbers of TSVs and interconnects required and achieved 45 % reduction in total TSV count.

9.3 3D Physical Design Methodology and CAD Support

As demands accelerate for increasing density, higher bandwidths, and lower power consumption, current semiconductor industries are looking up to 3D-ICs with TSVs as a possible solution to integrate a great deal of functionality into small form factors to improve performance and reduce cost. While there is great interest in this emerging technology, the 3D CAD tools development is still in its early phase. There are no common standard definitions available, 3D design, verification and test is still a challenge. From the 3D design and verification standpoint, the good news is that, we developed a 3D physical design methodology using Global Foundries

130 nm technology node modified to use Tezzaron's TSV technology. The design flow includes 2D design tools plus additional support programs to covers all areas of 3D design, including the design partitioning, merging multiple tiers (gds files) and design sign-off analysis. Our 3D physical design and verification methodology is automated by using a hierarchical RTL description code generator based on Tree-based FPGA architecture description and design constraints. A timing evaluation system based on Mentor's circuit simulator *Eldo* is attached to design module to accurately estimate the networks delays and TSV modeling. In addition, we also addressed the specific issues that 3D designers will encounter dealing with tools that are not specifically developed to meet their needs.

We designed and implemented a two-tier 3D Tree-based FPGA demonstrator with seven tree levels and 16K LBs using our new 3D physical design tool flow. A fast and accurate 3D thermal models which takes into account of the primitives and its characteristics to balance the heat produced across the die. Two different heat removal methods such as thermal design techniques and hardware-based techniques are implemented to improve the performance and reliability of the 3D FPGA chip. The 3D thermal model is embedded along with 3D physical design flow the evaluate the temperature profile of the chip. The 3D thermal ware design tool provides flexibility in using either design techniques or hardware based technique based of the design requirements as heat removal mechanism.

9.3.1 Interconnect Optimization Model

For an efficient FPGA design standpoint interconnect optimization model is essential, since FPGA is an interconnect dominated device. We developed Rent's Rule based interconnect optimization model based on 3D Tree-based FPGA router program implemented by clubbing a binary search algorithm in association *pathfinder* algorithm. The advantage of this model is that, it can optimize each level of the Tree-based interconnect separately as described in Chap. 6. The interconnect optimization model determine the TSV and routing resource requirements based on certain constraints, since the routing resource depopulation is a trade-off between speed and routability. The demonstrator test chip design and simulation shows that, using interconnect optimization model we removed 17,203 TSVs from vertically and 36,864 TSVs from horizontally partitioned two-tier 3D test chips with minimal impact on speed.

9.3.2 3D FPGA Architecture Exploration Tools and Technologies

In this book, we developed a complete architectural design and exploration CAD tool for 3D Tree based homogeneous and heterogeneous FPGA. Some of them are extended from 2D tools developed for previous work [1, 2]. The 3D design and interconnect network partitioning took advantage of the availability local communication to route signals and thereby reduce path delay of the circuits implemented. The horizontal partitioning of the interconnect network place all LBs and local hierarchical communication is tier 1 of the 3D design to focus on routing more signals using local communication and thereby discourage the router to use TSVs between tier 0 and tier 1. The performance efficiency of two-tier 3D Tree-based homogeneous and heterogeneous FPGA evaluated using the exploration tools developed for this work. The 3D Tree-based FPGA performed 1.5 times better compared to 3D Mesh-based FPGA with identical array size and logic density. We also analyzed the impact cluster arity and LUT size variation. Most of the modern FPGA has different cluster and LUT size. For 3D Tree based FPGA, we found LUT size 4–5 and cluster size 4–5 provides the best speed and area efficiency.

The idea of of using 3-dimension to design FPGAs is not new. Many researches and industrial start-ups demonstrated the creation of 3D-FPGAs by stacking 2D FPGAs and connecting them either with solder bumps or TSVs. The latest entry into FPGA market is the Xlinx's silicon interposer-based multi-FPGA technology is interesting, because it enables the creation of large FPGAs composed of small dies and very large FPGAs, with higher logic density than one can achieve with a single die using monolithic semiconductor technology. The Mesh-of-Tree (MoT)-based FPGA architecture described in Chap. 3 is an economical and efficient architecture to design and manufacture true 3D and 2.5D interposer-based multi-FPGAs. By unifying the good qualities of Tree- and Mesh-based FPGA architectures, we can design a true 3D FPGA with two or more tires and 2.5D interposer-based multi-FPGA using MoT-based FPGA dies by adjusting the long wire span and cut line position. Though not discussed in this book, we developed full set CAD tools for design and exploration for MoT-based FPGAs. The design validation and delay estimation were performed using a collection of large benchmarks selected from different sources. The 3D two-tier MoT-based FPGA improve area by 41 % and reduce delay by 54.26 %, which is very promising for high performance 3D FPGAs. The data from Xlinx's 2.5D interposer-based multi-FPGA shows 77 % of inter-FPGA wires are cut and various design version have different numbers of *cut*. We verified the delay variation of MoT-based 2.5D multi-FPGA with inter-FPGA wires removed in the range from 20–80 % and the we shows the delay increased by 3–4 nS. Our interposer-based multi-FPGA has 3 cuts and to improve the routability, we increase the W by 48 %. However in Mesh-based 2.5D architecture, it was increased to 125 % for same number of *cuts*.

9.4 Directions for Future Work

While the research work presented in this book is an important step in the development of 3D FPGAs and CAD tools that can support the design and implementation of modern FPGA architectures, there is much work that remains. FPGA technology began in the mid-1980s as an alternative to the popular ASIC technology of that time it was Gate Array (GA). During the 1990s, the Gate Array ASIC technology lost its appeal as more sophisticated ASIC technologies came to the fore, and the 20B$ Gate Array market shrunk dramatically until it effectively ceased to exist. Analysts expected that this would have a dramatic positive impact on the FPGA market, which did grow to some extent, but far less than everyone's expectations. We believe that the stagnation of FPGA growth is mostly due to the inefficiency of FPGA technology. Most FPGAs use SRAM as the programming or switch technology. Interconnects are the dominating resource in modern designs. Within an SRAM-based FPGA, the programming of interconnects is implemented by an SRAM cell that controls a pass-transistor driver. The research results from academia and semiconductor industry indicate that the cell area overhead for the SRAM-based programmable interconnect is over 30 times when compared to a via; and this does not include the additional circuit overhead area needed to program and control the SRAM-based routing resources. We found that with all the improvements in area and speed using 3D technology for circuits implemented purely using the LUT based logic elements, an FPGA is still has a much larger form factor and between 3–4 times slower on average than a standard-cell implementation. This high programmability overhead suggests that many of the current ASIC designs cannot be replaced by their FPGA equivalents and its imperative the FPGA architecture and design needs an overhaul change. Here we give few new research directions to improve the 3D FPGA technology described in this book to cover the gap between FPGA and ASIC and to make FPGA technology competent enough to reduce the exorbitant cost of design and manufacturing.

9.4.1 Technology Research

It is critical for both performance of the current architectures and invention of new ones, that a high-density interconnect be available. Specifically, we proposed few new high-density 3D Tree-based interconnect architectures to improve the performance and logic density of FPGA based systems. The main issues engineers trying to fix is the interconnect pitch must be within the order of magnitude of the minimum feature size. We also propose few technology level improvements as follows.

9.4.2 Alternative Memory Technology

As described in previous section, an alternative memory technology like CBRAM (Conductive Bridging Random Access Memories) [3] or RRAM (Resistive Random Access Memories) [4] instead of Six transistor SRAM would be better choice to reduce programmable interconnect overhead. There are few research reports academia shows significant improvement in the total area. Recently few FPGA manufactures decides to use 22-nm Tri-Gate CMOS process to design and implement high density FPGAs. It would be interesting to do a performance evaluation of 3D

Tree-based and MoT-based FPGAs using those alternative memory technologies and new 22 nm Tri-Gate CMOS process [5].

9.4.3 Monolithic 3D-FPGA

Recent results from many research institutes and 3D start-ups shows interesting data from 3D monolithic integration [6], which is considered as the only available option for applications requiring connections at the transistor scale. We plan to extend this work to also examine the impact of speed, power consumption and area improvement by using 3D monolithic stacking technology or native 3D integration methodology. This approach will further reduce the wire length and thereby improve performance of the FPGA based systems. Since the two-tier design is done in such a way to stack almost 80 % of the the programming overhead (tier 0) of Tree-based FPGA on top of logic blocks (tier 1) and interconnected using TSVs based multi-layer stacking technology. In the case of monolithic stacking the interconnect layers between programming overhead and logic blocks will be implemented in a state-of-the-art CMOS technology. This design and implementation methodology provide additional flexibility to improve logic density, speed and reduce power consumption and silicon area and thereby high density FPGAs can be designed and manufactured to meet the needs and requirements of high volume circuit design and prototyping. However the main challenge in this approach is to balance the density of TSVs to that of the via density in the CMOS technology used to implement logic and interconnect layers.

9.4.4 3D Hybrid FPGA (3D-HFPGA): CNT Based FPGA Interconnect

The major performance and power bottleneck of the FPGA is the programmable interconnects and routing elements inside the FPGA, which have been found to account for up to 80 % of the total delay and up to 85 % of the total power consumption when both local and global interconnects are considered. The integration of 3D FPGA with nano-materials and devices to establish high performance FPGA architecture may sheds new light on designing future programmable device. Carbon nanotubes (CNTs), nanowires, and other molecular electronic devices have demonstrated strong characteristics on improving speed, density and power consumption of semiconductor circuits. The combination of these two leading technologies shows a great potential for innovation and technology breakthroughs [7]. The Tree-based FPGA and MoT-based FPGA has tremendous potential to improve its performance, area and power consumption by using a combination of 3D integration and nanoelectronic devices. We believe it is an interesting idea to develop design and implementation technologies to demonstrate the performance of 3D hybrid FPGA (3D-hFPGA).

9.4.5 Mesh-of-Tree-based Embedded FPGA

Mesh-of-Tree-based FPGA architecture presented in this book is efficient scalable architecture suitable for the design and implementation 3D FPGAs in terms of smaller form factor and higher performance. FPGAs have been used in many applications to achieve orders-of-magnitude improvement in absolute performance and energy efficiency relative to conventional microprocessors. Even there are questions like *What if 10 % of a GPP/CPU Cache Area were FPGA?*. However recent research reports about the current Mesh-based FPGA architecture shows lack of essential abstraction that one comes to expect in a general purpose computer. MoT-based FPGA architecture presented in this dissertation is an efficient planar architecture suitable for the design and implementation embedded FPGAs. It would be an interesting idea to develop a design and exploration setup for MoT-based FPGA to study the advantages of MoT being an embedded FPGA in computing systems.

9.4.6 3D FPGA CAD Tools

Developing a design and implementation flow for 3D-FPGAs is a complicated task with many ramifications [8]. Design methodologies at the front-end and mature manufacturing technologies at the back-end are required to effectively facilitate large scale manufacturing of 3D systems. Several second-order digital optimization problems in 3D integration have yet to be solved. Optimizations in the space of 3D detailed routing likely exist that are not covered thermal performance during routing is a potential opportunity. There are many interesting questions for 3D and 2.5D Tree-based FPGA architectures still remains to be answered. Introducing an interconnect partitioning tool would figure number one item in our agenda for 3D FPGA CAD tool development. This methodology will help us to understand the wire-length optimization and placement of *break-point* before the development of physical design. Since the Tree-based FPGA layout is not scalable, it is important to have a predefined delay optimization process built in to determine the location of the *break-point*. Another additional feature for the CAD flow of 2.5D interposer-based MoTs need to be improved to check the routability automatically based on *number of cuts* and *% of inter-FPGA wires cut*. This will help the designer to set the W required for any given design specification.

References

1. Z. Marrakchi, H. Mrabet, C. Masson, H. Mehrez, Mesh of tree: unifying mesh and MFPGA for better device performances, in *NOCS-07*, pp. 243–252, 2007
2. Z. Marrakchi, H. Mrabet, U. Farooq, H. Mehrez, FPGA Interconnect topologies exploration. Int. J. Reconfig. Comput. 1–13 (2009)

3. S. Onkaraiah, O. Turkyilmaz, M. Reyboz, F. Clermidy, A hybrid CBRAM/CMOS look-up-table structure for improving performance efficiency of field-programmable-gate-array, in *IEEE International Symposium on Circuits and Systems (ISCAS)*, pp. 2440–2443, May 2013
4. Y.-C. Chen, W. Wang, H. Li, W. Zhang, Non-volatile 3D Stacking RRAM-based FPGA, in *22nd International Conference on Field Programmable Logic and Applications (FPL)*, pp. 367–372, Aug 2012
5. Altera White Papers, *The breakthrough advantage for FPGAs with tri-gate technology*, http://www.altera.com/literature/wp/wp-01201-fpga-tri-gate-technology.pdf (Altera Corporation USA, June 2013)
6. T. Naito, T. Ishida, T. Onoduka, M. Nishigoori, T. Nakayama, Y. Ueno, Y. Ishimoto, A. Suzuki, W. Chung, R. Madurawe, S. Wu, S. Ikeda, H. Oyamatsu, World's first monolithic 3D-FPGA with TFT SRAM over 90 nm 9 layer Cu CMOS, in *IEEE Symposium on VLSI Technology (VLSIT)*, pp. 219–220, June 2010
7. S. Eachempati, A. Nieuwoudt, A. Gayasen, N. Vijaykrishnan, Y. Massoud, Assessing carbon nanotube bundle interconnect for future FPGA architectures, in *Proceedings of the Conference on Design, Automation and Test in Europe, DATE'07*, pp. 307–312, 2007
8. N. Miyamoto, Y. Matsumoto, H. Koike, T. Matsumura, K. Osada, Y. Nakagawa, T. Ohmi, Development of a CAD tool for 3D-FPGAs, in *IEEE International 3D Systems Integration Conference (3DIC)*, pp. 1–6, Nov 2010

Appendix A
FPGA CAD Tool: 3D Homogeneous Tree-Based FPGA Architecture and Design Space Exploration

A.1 Architecture Description of Homogeneous Tree-Based Architecture

This section describe the prepration of input files to run the 3D homogeneous Tree-based FPGA architecture and design space exploration tool. This tool helps us the understand the partitioning, placment and routing of the multi-tier 3D FPGAs. The architecture specification of homogeneous 3D Tree-based FPGA architecture definition is presented in Table A.1. The variable nb_levels defined the number of levels in the Tree and *break_point* defines the break point for horizontal and vertical network partitioning methods. In this example the *break_point* is set for horizontal partitioning.

The tools allows to run placement and routing using command line options. The routing tool take the architecture definition, netlist file, placement file and timing file as inputs and compute the routability, area and speed of FPGA based on the type of benchmark used. The routing tools also computes the total number of switches, multiplexers and SRAM required for each benchmark application. A typical example shows below.

nb_levels = 8
timing_info levels 8
Parsing file : bench=pdc, cluster4, lut4,
arch=$4 \times 4 \times 4 \times 4 \times 4 \times 4\_rent4 \times 4 \times 4 \times 4 \times 4.desc$
build Tree-based FPGA ...
arch_param description ...
nb_levels 7
→ level 0 nb_slaves 4 nb_inputs 16 nb_outputs 4
→ level 1 nb_slaves 4 nb_inputs 64 nb_outputs 16
→ level 2 nb_slaves 4 nb_inputs 256 nb_outputs 64
→ level 3 nb_slaves 4 nb_inputs 1024 nb_outputs 256
→ level 4 nb_slaves 4 nb_inputs 4096 nb_outputs 1024

V. Pangracious et al., *Three-Dimensional Design Methodologies for Tree-based FPGA Architecture*, Lecture Notes in Electrical Engineering 350, DOI 10.1007/978-3-319-19174-4

Table A.1 Architecture description file parameters of tree-based homogeneous architecture

nb_levels	7						
.level	0	nb_slaves	4	nb_inputs	16	nb_outputs	4
.level	1	nb_slaves	4	nb_inputs	64	nb_outputs	16
.level	2	nb_slaves	4	nb_inputs	256	nb_outputs	64
.level	3	nb_slaves	4	nb_inputs	1024	nb_outputs	256
.break_Point	Level_opti	TSV_count					
.level	4	nb_slaves	4	nb_inputs	4096	nb_outputs	1024
.level	5	nb_slaves	4	nb_inputs	16384	nb_outputs	4096
.level	6	nb_slaves	4	nb_inputs	0	nb_outputs	0
.basic ble	nb_inputs	4	nb_outputs	1			
.basic in_iobs	level	1	number	1			
.basic out_iobs	level	1	number	1			
.word_width	8						

→ level 5 nb_slaves 4 nb_inputs 16384 nb_outputs 4096
→ level 6 nb_slaves 4 nb_inputs 0 nb_outputs 0
→ basic nb_in_iobs 1 nb_out_iobs 1 nb_inputs 4
In level 0 4096 UMSBs 4 : 4 → 16384 Mux 4
In level 1 4096 UMSBs 5 : 4 → 16384 Mux 5
In level 2 4096 UMSBs 4 : 4 → 16384 Mux 4
In level 3 4096 UMSBs 4 : 4 → 16384 Mux 4
In level 4 4096 UMSBs 4 : 4 → 16384 Mux 4
In level 5 4096 UMSBs 4 : 4 → 16384 Mux 4
In level 6 4096 UMSBs 4 : 4 → 16384 Mux 4
totalWiresDMSBs = 20 = 4 + 16
In level 0 16384 DMSBs 5 : 4 → 65536 Mux 5
totalWiresDMSBs = 80 = 16 + 64
In level 1 16384 DMSBs 5 : 5 → 81920 Mux 5
totalWiresDMSBs = 320 = 64 + 256
In level 2 16384 DMSBs 5 : 4 → 65536 Mux 5
totalWiresDMSBs = 1280 = 256 + 1024
In level 3 16384 DMSBs 5 : 4 → 65536 Mux 5
totalWiresDMSBs = 5120 = 1024 + 4096
In level 4 16384 DMSBs 5 : 4 → 65536 Mux 5
totalWiresDMSBs = 20480 = 4096 + 16384
In level 5 16384 DMSBs 5 : 4 → 65536 Mux 5
totalWiresDMSBs = 16384 = 16384 + 0
In level 6 16384 DMSBs 1 : 4 → 65536 Mux 1
Out IOBs clusters 1024 MSB 16:1 → 1024 MUX 16:1

LUT area = 66000 lambda 2
number of LUTs = 16384(mux 16:1) (mux 2:1) Flip-Flop
total Logic Gates = 6.35904e + 06
Logic area 1.07155e + 06 x 1000 λ^2
Upward network area 1177500 x 1000 λ^2
Downward network area 5567250 x 1000 λ^2
total_mux2_number = 2014208
total switches number = 2605056
total sram number = 1757184
total buffers number = 590848
total area = 7.83055e + 06 $\times$ 1000 λ^2

The exploration tool output data presented is useful to compute the total area and power consumption of the FPGA chip. For 3D multi-tier test chip analysis the *break-point* for the interconnect network partitioning should also be specified to optimize the number of I/O pins required.

Appendix B
FPGA CAD Tool: 3D Heterogeneous Tree-Based FPGA Exploration

B.1 Architecture Description of Heterogeneous Tree-Based Architecture

Different architecture parameters of the heterogeneous tree-based FPGA architecture are defined using an architecture description file. Some of these parameters are shown in Table B.1. The parameter Nb_Levels defines the total number of levels of the architecture. Nb_Block_Types parameter defines the total number of types that are supported by the architecture. By default, a tree-based architecture supports two types of blocks which are logic blocks (CLBs or soft blocks) and I/Os. For heterogeneous architectures, however, the types of blocks may vary depending on the netlist requirements that are being implemented on the architecture. The architecture is quite flexible in this sense and it can support any number of block types that can be placed at different levels of hierarchy in order to have a best design fit. In our description mechanism, the architecture description starts with the specification of I/O blocks and once it is done, the rest of the architecture is defined repeatedly using the parameters of lines 5 to 10 of Table B.1. These parameters include level number being defined, number of sub-cluster types supported by each cluster, number of sub-clusters contained in each cluster and number of inputs/outputs of the cluster. Definition of clusters starts from bottom level of the architecture and it goes to top until all the levels of the architecture are specified. Once cluster definition is over, binary search parameter is either set to be true or false. If binary search parameter is false, no architecture optimization is performed and the netlist is routed using the given cluster bandwidth. However if this parameter is true, then an architecture optimization is performed using the specified optimization approach which can be bottom_up, top_down or random.

Once cluster definition of all the levels is over, different types of blocks that are supported by the architecture can be defined using Define_Block parameter. Different parameters that are used for the definition of a block are shown in Table B.2. In a tree-based architecture, definition of a block starts with the name of the block. The

V. Pangracious et al., *Three-Dimensional Design Methodologies for Tree-based FPGA Architecture*, Lecture Notes in Electrical Engineering 350, DOI 10.1007/978-3-319-19174-4

Table B.1 Architecture description file parameters of tree-based architecture

	Name	Description
1.	Nb_Levels	Total number of levels in the architecture
2.	Nb_Block_Types	Number of block types that are supported by the architecture
3.	In_Blocks	The level of the cluster and the number of inputs per cluster of the input block
4.	Out_Blocks	The level of the cluster and the number of outputs per cluster of the output block
5.	Level	The level ℓ of the architecture
6.	Nb_Cluster_Type	Number of sub-cluster types supported by a cluster of level ℓ
7.	Arity	Number of sub-clusters of each type supported by a cluster of level ℓ
8.	Nb_Inputs_Per_Cluster	Number of inputs per cluster of each type
9.	Nb_Outputs_Per_Cluster	Number of outputs per cluster type
10.	End_Level	Completes the definition of level ℓ
11.	Optimization	Binary search flag set either true or false
12.	Optimization_approach	Specified as either bottom_up, top_down or random
13.	Define_Block blk	Block definition (See Table B.2)

Table B.2 Block definition in tree-based architecture

Definition	Description
Define_Block	
Block_Name	Name of the block
Area	Area of the block
Nb_Inputs	Number of inputs of the block
Nb_Outputs	Number of outputs of the block
Level_Number	Level number where the block is located
Arity	Number of blocks per cluster
Pin_Input	Name and the class number of input pins of the block
Pin_Output	Name and the class number of output pins of the block
End_Define_Block	

parameter *Area* gives the area of the block which is later used for the area calculation of the architecture. Other parameters include the number of input/output pins, the level where the block is located, the arity (i.e. no of blocks per cluster) of the block and the definition of its input and output pins. While defining I/O pins of a particular block (logic-block or a hard-block), unique class number are assigned to each block pin to ensure the appropriate routing of the netlist that is mapped on the architecture. The 3D heterogeneous Tree-based FPGA architecture is defined as follows.

B.1.1 Architecture Definition of Heterogeneous Tree-Based Architecture

nx 20
ny 16
hardware_portions false
horizontal_portions 2
vertical_portions 1
fpga true
t-driven Yes/No
optimized Yes/No
explore Yes/No
optimization_algo bottom_up: architecture optimization program selection nb_levels: 7 Tree level
nb_block_types 3
in_iobs level 1 number 5
out_iobs level 1 number 5
end_device

architecture_description:
.level: 0
Arity: 4
nb_cluster_type: 1
new_cluster_type:
nb_inputs_per_cluster: 16
nb_outputs_per_cluster: 4
end_cluster_type:
end_level: 0 First Level

.level: 1
Arity: 4
nb_cluster_type: 2
new_cluster_type:
nb_inputs_per_cluster: 64
nb_outputs_per_cluster: 16
end_cluster_type:
new_cluster_type:
nb_inputs_per_cluster: 0
nb_outputs_per_cluster: 0
end_cluster_type:
end_level: 1

.level: 2
Arity: 4

nb_cluster_type: 2
new_cluster_type:
nb_inputs_per_cluster: 256
nb_outputs_per_cluster: 64
end_cluster_type:
new_cluster_type:
nb_inputs_per_cluster: 0
nb_outputs_per_cluster: 0
end_cluster_type:
end_level: 2

.level: 3
Arity: 4
nb_cluster_type: 1
new_cluster_type:
nb_inputs_per_cluster: 1024
nb_outputs_per_cluster: 256
end_cluster_type:
end_level: 3

.level: Break_Point TSV_Count

.level: 4
Arity: 4
nb_cluster_type: 1
new_cluster_type: nb_inputs_per_cluster: 4096
nb_outputs_per_cluster: 1024
end_cluster_type:
new_cluster_type:
nb_inputs_per_cluster: 36
nb_outputs_per_cluster: 37
end_cluster_type:
end_level: 4

.level: 5
Arity: 4
nb_cluster_type: 1
new_cluster_type:
nb_inputs_per_cluster: 16384
nb_outputs_per_cluster: 4096
end_cluster_type:
end_level: 5

.level: 6
Arity: 4
nb_cluster_type: 1
new_cluster_type:
nb_inputs_per_cluster: 0
nb_outputs_per_cluster: 0
end_cluster_type:
end_level: 6
end_architecture_description:

block ble
area 66000
nb_inputs 4
nb_outputs 1
level 0
number 4
total_blocks 256
x_slots 1
y_slots 1
pin_input i0 0
pin_input i1 1
pin_input i2 2
pin_input i3 3
pin_output q0 4
end_block

block mult_36
area 2516000
nb_inputs 36
nb_outputs 37
total_inputs 36
total_outputs 37
level 4
number 1
total_blocks 4
x_slots 4
y_slots 4
pin_input a0 0, pin_input a1 1, pin_input a2 2, pin_input a3 3
pin_input a4 4, pin_input a5 5, pin_input a6 6, pin_input a7 7
pin_input a8 8, pin_input a9 9
pin_input a10 10, pin_input a11 11, pin_input a12 12, pin_input a13 13
pin_input a14 14, pin_input a15 15, pin_input a16 16, pin_input a17 17
pin_input b0 18, pin_input b1 19, pin_input b2 20, pin_input b3 21
pin_input b4 22, pin_input b5 23, pin_input b6 24, pin_input b7 25

```
pin_input b8 26, pin_input b9 27
pin_input b10 28, pin_input b11 29, pin_input b12 30, pin_input b13 31
pin_input b14 32, pin_input b15 33, pin_input b16 34, pin_input b17 35
pin_output q0 36, pin_output q1 37, pin_output q2 38, pin_output q3 39
pin_output q4 40, pin_output q5 41, pin_output q6 42, pin_output q7 43
pin_output q8 44, pin_output q9 45, pin_output q10 46, pin_output q11 47
pin_output q12 48, pin_output q13 49, pin_output q14 50, pin_output q15 51
pin_output q16 52, pin_output q17 53, pin_output q18 54, pin_output q19 55
pin_output q20 56, pin_output q21 57
pin_output q22 58, pin_output q23 59
pin_output q24 60, pin_output q25 61
pin_output q26 62, pin_output q27 63
pin_output q28 64, pin_output q29 65
pin_output q30 66, pin_output q31 67
pin_output q32 68, pin_output q33 69
pin_output q34 70, pin_output q35 71
pin_output q36 72
end_block
```

Appendix C
FPGA CAD Tool: 3D MoT-Based FPGA Exploration

C.1 Architecture Description of Mesh-of-Tree FPGA Architecture

nx = 36
ny = 36
in_rate = 8
out_rate = 8
lut size = 4
network width = 38 # cutline horizontal or vertical: The *cutline can be place in horizontal or vertical direction based on the FPGA architecture and design*
constrained_channels horz
In this example we used horizontal *cutline*, for which the specification at the command line constrained_channels horz
constrained_X 16
constrained_channels = horz, constrained_Y the FPGA is cut at location X=16
constrained_Y 16
constrained_width_x 8
constrained_width_y 8
As we discussed in Chap. 3, for 2.5D FPGA we have the flexibility to remove interdire FPGA channels wires along the *cutline*. The number provided at command line constrained_width_x and constrained_width_y will be kept and rest of the wires will be removed.
archCluster size = 8
#arch cluster
archCluster inputs = 24
#archClusterOutputsNB : 1 parametrable, 0 non parametrable
archClusterOutputsNBparam 1
archCluster outputs = 8
#netlist clb

V. Pangracious et al., *Three-Dimensional Design Methodologies for Tree-based FPGA Architecture*, Lecture Notes in Electrical Engineering 350, DOI 10.1007/978-3-319-19174-4

netClb inputs = 12
longWire horizontal 2 continue true ratio 0.4
longWire vertical 2 continue true ratio 0.4
long wire segments: In this example the value of a is set to 0.4, which means we have 40 % of additional long wire with a span of 2 is set in addition to 38 horizontal channels. Here the value of $W = 38$ and $a = 0.4$. So the number of long wire segments equal to $a \times W$, which means = 0.4×38, which is ≈ 16 channels. So the total number of channels for any switch block is equal to $38 + 16 = 54$. The architecture file can be expanded to include multiple *cutline* and apply channel constraints independently.

Appendix D
3D Tree-Based FPGA Thermal Modeling

D.1 3D Thermal Model: Description of Multi-tier FPGA Dies

The 3D thermal model is designed to work along with the physical design tools and has the capability to read the layout parameters like block level geometrical features and power values. As described in Chap. 9, we have coded additional capabilities to include special hotspot zones for heat transfer. The Table D.1 shows the format of special hotspot zones. The effective thermal conductivity is calculated based the area occupied by the TSVs, type of materials used (copper or tungsten), liner material.

D.1.1 3D File Format

File Format:
Layer Number This variable specify this particular occupy which position inside the 3D stack
Transverse heat flow Y/N ?
Lateral heat flow Y/N ?: The direction of heat flow can laos mentioned based on the type of components in the active layer. For example, The heat transfer in an adhesive layer is considered to be perpendicular to the device plane and no heat flux occurs along the adhesive layer due to its low thermal conductivity. The heat transfer in TSVs is considered as one dimensional and perpendicular to the device plane. The heat flow inside the 3D stack is diffusive in nature.
Power Dissipation Y/N?: The power sources are provided using another input file along with the command line option. This variable takes the inputs from the power source file.
Specific heat capacity in $J/(m^3K)$, this variable provide the specific heat capacity of active layer.
Resistivity in (m K)/W

V. Pangracious et al., *Three-Dimensional Design Methodologies for Tree-based FPGA Architecture*, Lecture Notes in Electrical Engineering 350, DOI 10.1007/978-3-319-19174-4

Table D.1 Tier 0, special hotspot zone definition

Unit-name	3D Layer	Width	Height	Left-X_dir	Bottom-Y_dir	TTSV count
Zone0	Layer_2	0.000750	0.002000	0.000000	0.000000	3
Zone1	Layer_1	0.000500	0.002000	0.000750	0.000000	8
Zone2	Layer_2	0.000750	0.002000	0.001250	0.000000	2

Table D.2 Tier 1, layout description of tree-based homogeneous architecture: part I

Unit-name	Width	Height	Left-X_dir	Bottom-Y_dir
b0	0.000250	0.000250	0.000000	0.000000
l0a	0.000250	0.000350	0.000000	0.000250
b1	0.000250	0.000250	0.000000	0.000600
l1a	0.000250	0.000850	0.000250	0.000000
b2	0.000250	0.000250	0.000500	0.000000
l0b	0.000250	0.000350	0.000500	0.000250
b3	0.000250	0.000250	0.000500	0.000600
l3a	0.000500	0.002000	0.000750	0.000000
b4	0.000250	0.000250	0.001250	0.000000
l0c	0.000250	0.000350	0.001250	0.000250
b5	0.000250	0.000250	0.001250	0.000600
l1b	0.000250	0.000850	0.001500	0.000000
b6	0.000250	0.000250	0.001750	0.000000
l0d	0.000250	0.000350	0.001750	0.000250
b7	0.000250	0.000250	0.001750	0.000600
l2a	0.000750	0.000300	0.000000	0.000850
l2b	0.000750	0.000300	0.001250	0.000850
b8	0.000250	0.000250	0.000000	0.001150
l0e	0.000250	0.000350	0.000000	0.001400
b9	0.000250	0.000250	0.000000	0.001750
l1c	0.000250	0.000850	0.000250	0.001150
b10	0.000250	0.000250	0.000500	0.001150
l0f	0.000250	0.000350	0.000500	0.001400
b11	0.000250	0.000250	0.000500	0.001750

Thickness in μ m or mm or in meters The exact unites of length and width is depend of the geometrical output from the layout editor. The simulator has the capability to set the unites based on the unit of the block dimensions.

Floorplan file: The floorplan file provides information about the placements of different blocks and an example shown in Tables D.2 and D.3

Table D.3 Tier 1, layout description of tree-based homogeneous architecture: part II

Unit-name	Width	Height	Left-X_dir	Bottom-Y_dir
b12	0.000250	0.000250	0.001250	0.001150
l0g	0.000250	0.000350	0.001250	0.001400
b13	0.000250	0.000250	0.001250	0.001750
l1d	0.000250	0.000850	0.001500	0.001150
b14	0.000250	0.000250	0.001750	0.001150
l0h	0.000250	0.000350	0.001750	0.001400
b15	0.000250	0.000250	0.001750	0.001750

D.1.1.1 Output File from 3D Thermal Model

The 3D thermal model provides 2D and 3D thermal profile of the chip. The simulation mode (2D or 3D) can be changed using command line input. The thermal model is embedded along with physical design tools to evaluate the thermal profile of 2D and 3D designs and the temperature profile can be used to decide the changes required in floorplan to optimize thermal profile of the designs. The Tables D.4 and D.5 shows the layer wise temperature profile of each blocks in the designs. The design model we used is a hierarchical and the design is divided into many blocks and combined together during the design phase. The block level thermal analysis is very useful for such designs.

Table D.4 3D thermal simulator output file format: 3D thermal profile of the chip part I

Active layer/Unite name	Temperature (K)	Active layer/Unite name	Temperature (K)
layer_0_b0	364.20	layer_0_l0a	364.69
layer_0_b1	364.62	layer_0_l1a	365.45
layer_0_b2	366.57	layer_0_l0b	366.43
layer_0_b3	367.04	layer_0_l3a	370.46
layer_0_b4	367.16	layer_0_l0c	367.55
layer_0_b5	367.39	layer_0_l1b	366.04
layer_0_b6	364.58	layer_0_l0d	365.03
layer_0_b7	364.76	layer_0_l2a	365.80
layer_0_l2b	365.79	layer_0_b8	364.82
layer_0_l0e	365.16	layer_0_b9	364.58
layer_0_l1c	365.96	layer_0_b10	367.43
layer_0_l0f	367.61	layer_0_b11	367.23
layer_0_b12	367.65	layer_0_l0g	367.98
layer_0_b13	367.51	layer_0_l1d	366.35
layer_0_b14	364.91	layer_0_l0h	365.33
layer_0_b15	364.86		

Table D.5 3D thermal simulator output file format: 3D thermal profile of the chip part II

Active layer/Unite name	Temperature (K)	Active layer/Unite name	Temperature (K)
layer_2_level4_Z1	362.60	layer_2_Level5_Z1	366.81
layer_2_Level6_Z1	362.61	layer_2_Level5	351.93
layer_2_Level4	354.64	layer_2_Level6	351.99
layer_2_Level4_BlockSB1	345.92	layer_2_Level5_BlockSB2	348.14
layer_2_Level4_BlockSB3	345.97		
layer_2_Unit1_Int	325.91	layer_2_Unit2_Int	326.48
layer_2_Unit3_Int	325.92		
Probe_node_0	320.30	Probe_node_1	320.30
Probe_node_2	320.31	Probe_node_3	320.31
Probe_node_4	320.28	Probe_node_5	320.28
Probe_node_6	320.28	Probe_node_7	320.28
Probe_node_8	319.75	Probe_node_9	319.75
Probe_node_10	319.76	Probe_node_11	319.76
hsink_Unit1	320.97	hsink_Unit2	321.04
hsink_Unit3	320.97		

The layer 1 of the 3D Tree-based FPGA contains SBs, HBs and programmable interconnects. To transfer heat, we also used special heat transfer zones as presented in Table D.5. The temperature estimated at different points inside the layer 2 of the 3D Tree-based FPGA chip is presented in Table D.5. One additional item we have in 3D thermal model is place probe point to monitor the temperature. This provision is introduced to monitor the temperature growth rate of certain devices placed in 3D chip dynamically. The Table D.5 shows few of those probe points and temperature values.

MIX
Papier aus verantwortungsvollen Quellen
Paper from responsible sources
FSC® C105338

If you have any concerns about our products,
you can contact us on
ProductSafety@springernature.com

In case Publisher is established outside the EU,
the EU authorized representative is:
Springer Nature Customer Service Center GmbH
Europaplatz 3, 69115 Heidelberg, Germany

Printed by Libri Plureos GmbH
in Hamburg, Germany